Finite Element Simulations Using
ANSYS

Esam M. Alawadhi

CRC Press
Taylor & Francis Group
Boca Raton London New York

CRC Press is an imprint of the
Taylor & Francis Group, an **informa** business

CRC Press
Taylor & Francis Group
6000 Broken Sound Parkway NW, Suite 300
Boca Raton, FL 33487-2742

First issued in paperback 2017

© 2010 by Taylor and Francis Group, LLC
CRC Press is an imprint of Taylor & Francis Group, an Informa business

No claim to original U.S. Government works

ISBN-13: 978-1-4398-0160-4 (hbk)
ISBN-13: 978-1-138-11183-7 (pbk)

Library of Congress Cataloging-in-Publication Data

Alawadhi, Esam M.
  Finite element simulations using ANSYS / Esam M. Alawadhi.
    p. cm.
  "A CRC title."
  Includes bibliographical references and index.
  ISBN 978-1-4398-0160-4 (hard back : alk. paper)
  1. Finite element method--Data processing. 2. ANSYS (Computer system) I. Title.

TA347.F5A37 2010
620.001'51825--dc22   2009019904

Visit the Taylor & Francis Web site at
http://www.taylorandfrancis.com

and the CRC Press Web site at
http://www.crcpress.com

Contents

Preface

Due to the complexity of modern-day problems in mechanical engineering, relying on pure theory or pure experiments is seldom practical. The use of engineering software is becoming prevalent among academics as well as practicing engineers. For a large class of engineering problems, especially meaningful ones, writing computer codes from scratch is seldom found in practice. The use of reputable, trustworthy software can save time, effort, and resources while still providing reliable results.

This book focuses on the use of ANSYS in solving practical engineering problems. ANSYS is extensively used in the design cycle by industry leaders in the United States and around the world. Additionally, ANSYS is available in computer laboratories in most renowned universities and institutes around the world. Courses such as computer aided design (CAD), modeling and simulation, and core design all utilize ANSYS as a vehicle for performing modern engineering analyses. Senior students frequently incorporate ANSYS in their design projects. Graduate-level finite element courses also use ANSYS as a complement to the theoretical treatment of the finite element method.

The book provides mechanical engineering students and practicing engineers with a fundamental knowledge of numerical simulation using ANSYS. It covers all disciplines in mechanical engineering: structure, solid mechanics, vibration, heat transfer, and fluid dynamics, with adequate background material to explain the physics behind the computations. It treats each physical phenomenon independently to enable readers to single out subjects or related chapters and study them as a self-contained unit. Instructors can liberally select appropriate chapters to be covered depending on the objectives of the course. For example, multiphysics analyses, such as structure–thermal or fluid–thermal analyses, are first explained theoretically, the equations governing the physical phenomena are derived, and then the modeling techniques are presented. Each chapter focuses on a single physical phenomenon, while the last chapter is devoted to multiphysics analyses and problems. The basic required knowledge of the finite element method relevant to each physical phenomenon is illustrated at the beginning of the respective chapter. The general theory of the finite element, however, is presented in a concise manner because the theory is well documented in other finite element books.

Each chapter contains a number of pictorially guided problems with appropriate screenshots constituting a step-by-step technique that is easy to follow. Practical end-of-chapter problems are provided to test the reader's understanding. Several practical, open-ended case studies are also included in the problem sections. Additionally, the book contains a number of complete tutorials on using ANSYS for real, practical problems. Because a finite element solution is greatly affected by the quality of the mesh, a separate chapter on mesh generation is included as a simple meshing guide, emphasizing the basics of the meshing techniques. The book is written in such a manner that it can easily be used for self-study. The main objective of this book is guiding the reader from the basic modeling requirements toward getting the correct and physically meaningful numerical result. Many of the sample problems, questions, and solved examples were used in CAD courses in many universities around the world. The topics covered are

1. Structural analysis
2. Solid mechanics and vibration
3. Steady-state and transient heat-transfer analysis
4. Fluid dynamics
5. Multiphysics simulations, including thermal–structure, thermal–fluid, and fluid–structure
6. Modeling and meshing guide

Undergraduate and graduate engineers can use this book as a part of their courses, either when studying the basics of applied finite elements or in mastering the practical tools of engineering modeling. Engineers in industry can use this book as a guide for better design and analysis of their products. In all mechanical engineering curricula, junior- and senior-level courses use some type of engineering modeling software, which is, in most cases, ANSYS. Senior students also use ANSYS in their design projects. Graduate-level finite element courses frequently use ANSYS to comple-ment the theoretical analysis of finite elements. The courses that use this book should be taken after an introduction to design courses, and basic thermal–fluid courses. Courses such as senior design can be taken after this course.

Acknowledgments

I profited greatly from discussion with faculty members and engineers at Kuwait University, particularly Professor Ahmed Yigit and Engineer Lotfi Guedouar.

1 Introduction

1.1 FINITE ELEMENT METHOD

The basic principles of the finite element method are simple. The first step in the finite element solution procedure is to divide the domain into elements, and this process is called discretization. The elements' distribution is called the mesh. The elements are connected at points called nodes. For example, consider a gear tooth, as shown in Figure 1.1. The region is divided into triangular elements with nodes at the corners.

After the region is discretized, the governing equations for each element must be established for the required physics. Material properties, such as thermal conductivity for thermal analysis, should be available. The elements' equations are assembled to obtain the global equation for the mesh, which describes the behavior of the body as a whole. Generally, the global governing equation has the following form

$$[K]\{A\} = \{B\}$$ (1.1)

where
  $[K]$ is called the stiffness matrix
  $\{A\}$ is the nodal degree-of-freedom, the displacements for structural analysis, or temperatures for thermal analysis
  $\{B\}$ is the nodal external force, forces for structural analysis, or heat flux for thermal analysis

The $[K]$ matrix is a singular matrix, and consequently it cannot be inverted.

Consider a one-dimensional bar with initial length L subjected to a tensile force at its ends, as shown in Figure 1.2. The cross-section area of the bar is A. The bar can be modeled with a single element with two nodes, i and j, as shown in Figure 1.2.

Assuming that the displacement of the bar, $d(x)$, varies linearly along the length of the bar, the expression of the displacement can be represented as

$$d(x) = a + bx$$ (1.2)

The displacement at node i and j are d_i and d_j, respectively. Then,

$$d_i(x) = a + bx_i$$ (1.3)

1

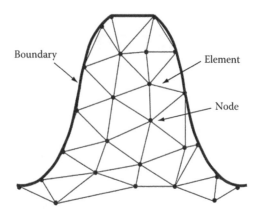

FIGURE 1.1   Finite element mesh of a gear tooth.

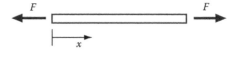

FIGURE 1.2   One-dimensional bar element.

$$d_j(x) = a + bx_j \tag{1.4}$$

where
   x_i is the x-coordinate for node i
   x_j is the x-coordinate for node j

Solving for a and b, it is found that

$$a = \frac{(d_i x_j - d_j x_i)}{L} \tag{1.5}$$

$$b = \frac{(d_i - d_j)}{L} \tag{1.6}$$

where L is the length of the element, $L = x_i - x_j$. Substituting a and b into the displacement equation (Equation 1.2) and after rearranging, the displacement function becomes

$$d(x) = \frac{x_j - x}{L} d_i + \frac{x - x_i}{L} d_j \tag{1.7}$$

or

$$d(x) = N_i d_i - N_j d_j \tag{1.8}$$

where N_i and N_j are called the shape functions of the element. When the bar is loaded, it will be in an equilibrium position. The sum of the strain energy (γ) and work (w) done by external force is the potential energy (π) of the bar. The potential energy at the equilibrium position must be minimized, and it is defined as

$$\pi = \gamma - w \tag{1.9}$$

For a single bar element, the strain energy stored in the bar is given by

$$\gamma = \int_{x_j}^{x_i} \frac{1}{2} \sigma \epsilon A \, dx \tag{1.10}$$

Since strain is related to stress by Young's modulus ($\sigma = E\epsilon$), the strain energy can be expressed as

$$\gamma = \frac{AE}{2} \int_{x_j}^{x_i} \epsilon^2 \, dx \tag{1.11}$$

The strain is equal to the elongation of the bar in the x-direction, as follows:

$$\epsilon = \frac{(d_j - d_i)}{L} \tag{1.12}$$

The strain energy in Equation 1.11 becomes

$$\gamma = \frac{AE}{2L} (d_j - d_i)^2 \tag{1.13}$$

In matrix form,

$$\gamma = \frac{AE}{2L} \begin{bmatrix} d_i & d_j \end{bmatrix} \begin{bmatrix} 1 & -1 \\ -1 & 1 \end{bmatrix} \begin{Bmatrix} d_i \\ d_j \end{Bmatrix} = \frac{1}{2} \{d\}^T [K] \{d\} \tag{1.14}$$

The work done by the applied forces at the nodes can be expressed as

$$w = d_i F_i + d_j F_j = \{d\}^T \{F\} \tag{1.15}$$

Hence, for a single bar element, the total potential energy becomes

$$\gamma = \frac{1}{2}\{d\}^T [K]\{d\} - \{d\}^T \{F\} \tag{1.16}$$

For the minimum potential energy, the displacement must be

$$\frac{\partial \gamma}{\partial\{d\}} = 0 \tag{1.17}$$

Therefore, Equation 1.16 becomes

$$[K]\{d\} - \{F\} = 0 \tag{1.18}$$

and therefore

$$\begin{Bmatrix} F_i \\ F_j \end{Bmatrix} = \frac{AE}{L} \begin{bmatrix} 1 & -1 \\ -1 & 1 \end{bmatrix} \begin{Bmatrix} u_i \\ u_j \end{Bmatrix} \tag{1.19}$$

The above derivation is valid only for one bar element. In practice, a model consists of many elements of different properties. The total potential energy of E number of elements is

$$\pi = \sum_{e=1}^{E} (\gamma^e - w) \tag{1.20}$$

Minimizing the above equation, it is found that

$$\sum_{e=1}^{E} [K^e]\{d\} - \{F\} = 0 \tag{1.21}$$

1.2   ELEMENT TYPES

Depending on the problem, the elements can have different shapes, such as lines, areas, or volumes. Figure 1.3 shows the basic element types. The line elements are used to model trusses made of spring, links, or beams. The area elements that could

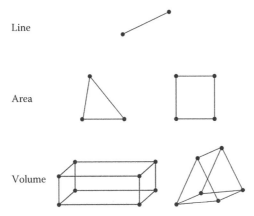

FIGURE 1.3 Basic element types.

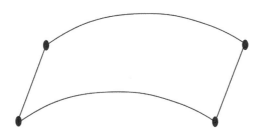

FIGURE 1.4 A shell element.

be rectangle or triangle are used to model two-dimensional solid areas, such as stress analysis for a plate, or fins for thermal analysis.

The volume elements are used to model three-dimensional bodies. Shells are special elements. They do not fall into either the area or the volume division. They are essentially two-dimensional in nature, but the area of the elements can be curved to model a three-dimensional surface. Figure 1.4 shows a rectangular shell element. This type of element is very effective for modeling very thin bodies such as cans under a stress.

1.3 SYMMETRIES IN MODELS

The discretization of a body is the first step in the finite element solution, where the body could be in two- or three-dimension space. The discretization should be in a certain way to avoid potential errors and to save time and effort. All structures in the real world are three-dimensional space. However, some effective approximations can be made to reduce the size of the computation domain. If the geometry and loads of a problem can be completely described in one plane, then the model

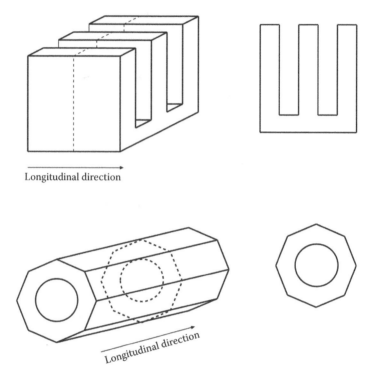

Longitudinal direction

Longitudinal direction

FIGURE 1.5   Two-dimensional plane representations of three-dimensional objects.

can be modeled in two-dimensional space. For example, consider the analysis of an electronic fin and a shaft, as shown in Figure 1.5. These objects are long, and if the loads are not varied in the longitudinal direction, then the physical characteristics do not vary significantly in the longitudinal direction. The two-dimensional assumption is valid anywhere in the object except at the ends.

There are four common types of symmetry encountered in an engineering problem:

1. Axial
2. Planar
3. Cyclic
4. Repetitive

In the axial symmetry cases, the variables distribution is constant in the circumferential direction. The distribution and loading are confined to only two directions, the radial and axial. In these problems, the axisymmetric elements must be used to mesh the model. Figure 1.6 shows an example of the axial symmetry for heat transfer problem of a pipe. The top and bottom surfaces are maintained at uniform temperatures, while the inner and outer surfaces are insulated. The boundary conditions should not be modified.

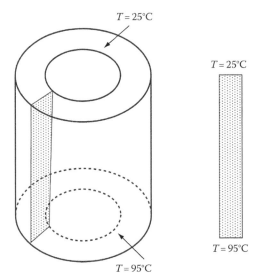

FIGURE 1.6   Type of symmetry: axial.

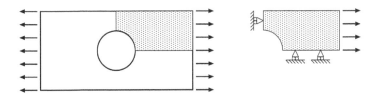

FIGURE 1.7   Type of symmetry: planar.

Consider a flat plate with a hole, as shown in Figure 1.7. A tensile pressure is applied at the left and right sides. It is only necessary to consider one-quarter of the problem. This type of symmetry is called the planar. The boundary condition should be modified at the surfaces of symmetry. For the example shown in Figure 1.7, the vertical line of symmetry should have zero displacement in the x-direction, and the horizontal line of symmetry should have a zero displacement in the y-direction.

The cyclic symmetry is similar to the planar symmetry except that it is described in a cylindrical rather than a rectangular coordinate system. The common problem found in practice is a washer under stress, as shown in Figure 1.8.

When the geometry and boundaries of the model are repeated in a particular direction, then the model has a repetitive symmetry. The repetitive symmetry is also called a periodic. For example, the simulation of a flow in a wavy channel can be easily modeled using the repetitive symmetry, as shown in Figure 1.9. Special periodic boundary conditions should be imposed. In addition, the size of a long beam with holes under a distributed pressure can be significantly reduced using the repetitive symmetry. In this case, the vertical line of symmetry should have a zero displacement in the x-direction, as shown in Figure 1.9.

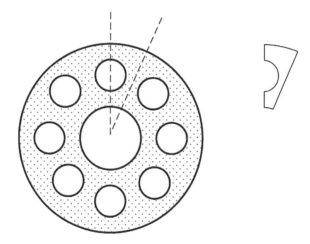

FIGURE 1.8 Type of symmetry: cyclic.

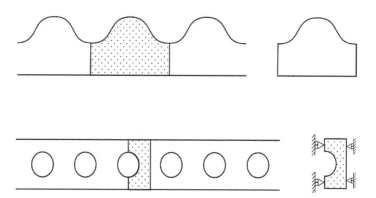

FIGURE 1.9 Type of symmetry: repetitive.

1.4 INTRODUCTION TO ANSYS®

ANSYS software is the most advanced package for single- and multiphysics simulations, offering enhanced tools and capabilities that enable engineers to complete their jobs in an efficient manner. ANSYS includes significant capabilities, expanding functionality, and integration with almost all CAD drawing software, such as Pro/ENGINEER, AutoCAD, and Solid Edge. In addition, ANSYS has the best-in-class solver technologies, an integrated coupled physics for complex simulations, integrated meshing technologies customizable for physics, and computational fluid dynamics (CFD).

ANSYS can solve problems in structural, thermal, fluid, acoustics, and multiphysics:

Structural

- Linear
- Geometric and material nonlinearities
- Contact
- Static
- Dynamic
- Transient, natural frequency, harmonic response, response spectrum, random vibration
- Buckling
- Topological optimization

Thermal

- Steady state or transient
- Conduction
- Convection
- Radiation
- Phase change

Fluids

- Steady state or transient
- Incompressible or compressible
- Laminar or turbulent
- Newtonian or non-Newtonian
- Free, forced, or mixed convection heat transfer
- Conjugate solid/fluid heat transfer
- Surface-to-surface radiation heat transfer
- Multiple species transport
- Free surface boundaries
- Fan models and distributed resistances
- Stationary or rotating reference frames

Acoustics

- Fully coupled fluid/structural
- Near- and far-field
- Harmonic, transient, and modal

Multiphysics

- Thermal–mechanical
- Thermal–electric
- Thermal–electric–structural
- Piezoelectric
- Piezoresistive
- Peltier effect

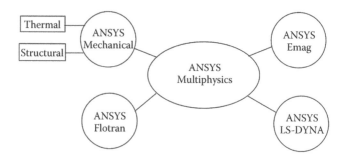

FIGURE 1.10 The ANSYS family.

- Seeback effect
- Thermocouple effect
- Electromechanical circuit simulator

ANSYS is not a single program, but it is a family of CAD programs (Figure 1.10). The ANSYS Multiphysics package consists of ANSYS/Emag for magnetic field analysis, ANSYS/Flotran for fluid dynamics, ANSYS/LS-DYNA for dynamics analysis, and ANSYS/Mechanical for structural and thermal analyses.

To start ANSYS, double click on the ANSYS icon or *Start > Programs > ANSYS > ANSYS Product Launcher.* The ANSYS Product Launcher window will show up. First, select the ANSYS Multiphysics. Then, enter the location for ANSYS files in the Working Directory. All ANSYS files will be stored in this directory including images and AVI files. The session name is entered in the Job Name. Finally, run ANSYS, as shown below:

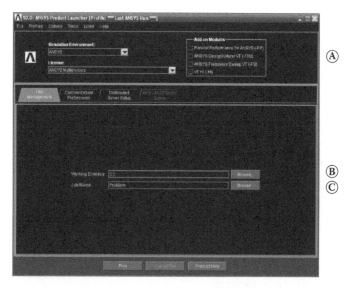

A Select multiphysics in License

B Change the working directory to C:/

C Change the initial job name to Problem

Run

Then the ANSYS windows will be shown with the ANSYS Output. The ANSYS window has the following components:

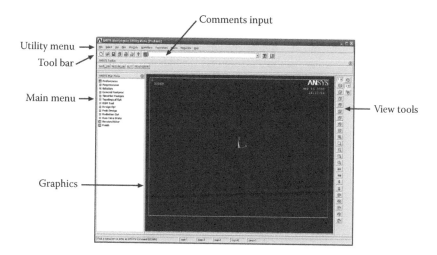

Comments input

Utility menu →

Tool bar →

Main menu →

Graphics →

View tools →

In addition, the ANSYS Output window will show up. The Output window dynamically lists important information during the preprocessor, solution, and postprocessor. Any warning in the ANSYS Output should be carefully considered to avoid unexpected errors.

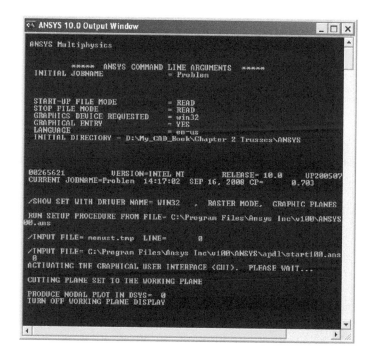

ANSYS Utility Menu

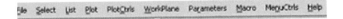

Within the Utility Menu, the file operations, list and plot items, and change display options can be done. In the Pull-down File Menu, the following tasks can be performed:

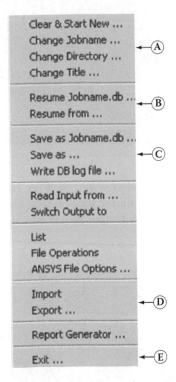

A Clear and starting a new job. This operation will not restart the log or the error files

B This task is for resuming a job, the database file should be available

C To save the current work

D To import geometry to ANSYS or to export an ANSYS model

E To exit ANSYS

The Pull-down List Menu is for listing the model components, such as key points, areas, and nodes. In addition, the properties of the material can be listed. The boundary conditions imposed on the model can also be listed.

The Pull-down plot Menu is similar to the List Menu. Plotting geometry's components, such as key points, can be performed in this menu. In the PlotCtrls Menu, printing the model in the ANSYS graphics, changing the style of the ANSYS graphics, or changing the quality of the graphics can be done. The Workplane is for the grids setup.

ANSYS Main Menu

Most of the ANSYS jobs are done in the Main Menu, from building the model to getting the results. In the Preprocessor, the material properties are assigned, real constants are specified, element type is selected, and model building and meshing tools are available. In the solution, the boundary conditions are imposed, and the solution setup parameters are specified. In the Postprocessor, the presentation of the ANSYS results is performed. List, plot results, and path operations can be performed. The three tasks are summarized as follows:

Preprocessor

1. Element type
2. Material properties
3. Real constants
4. Modeling
5. Meshing

Solution

1. Boundary condition
2. Solution setup parameters

Postprocessor

1. Plot results
2. List results
3. Path operation

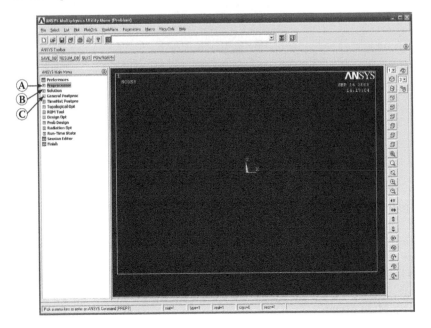

A Preprocessor

B Solution

C Postprocessor

2 Trusses

2.1 DEVELOPMENT OF BAR ELEMENTS

The derivation of the stiffness matrix for a bar element is applicable to the solution of pin-connected trusses. The bar element is assumed to have a constant cross-section area A, uniform modulus of elasticity E, and initial length L. The bar is subjected to tensile forces along the local axis that are applied at its ends. There are two coordinate systems: a local one $(\hat{x}, \hat{y})$ and a global one (x, y). The nodal degrees of freedom are the four local displacements: $\hat{d}_{1x}$, $\hat{d}_{1y}$, $\hat{d}_{2x}$, and $\hat{d}_{2y}$. The strain–displacement relationship is obtained from Hooke's law

$$\sigma_x = E\epsilon_x \tag{2.1}$$

where

$$\epsilon_x = \frac{d\hat{u}}{d\hat{x}} \tag{2.2}$$

$$A\sigma_x = T \tag{2.3}$$

$\hat{u}$ is the axial displacement in the $\hat{x}$-direction
T is the tensile force

Note that the bar element cannot sustain shear forces. Substituting σ_x and ϵ_x into Hooke's law:

$$\frac{d}{d\hat{x}}\left(AE\frac{d\hat{u}}{d\hat{x}}\right) = 0 \tag{2.4}$$

Assuming a linear displacement along the local x-axis of the bar, the displacement function can be written as

$$\hat{u} = a_1 + a_2\hat{x} \tag{2.5}$$

Now, the displacement function is expressed as a function of the nodal displacement $\hat{d}_{1x}$ and $\hat{d}_{2x}$, which can be achieved by evaluating $\hat{u}$ at the nodes, and solving for a_1 and a_2, as follows

$$\hat{u}(0) = \hat{d}_{1x} = a_1 \tag{2.6}$$

$$\hat{u}(L) = \hat{d}_{2x} = a_2 L + \hat{d}_{1x} \tag{2.7}$$

and solving for a_2

$$a_2 = \frac{\hat{d}_{2x} - \hat{d}_{1x}}{L} \tag{2.8}$$

The displacement function becomes

$$\hat{u} = \left(\frac{\hat{d}_{2x} - \hat{d}_{1x}}{L} \right) \hat{x} + \hat{d}_{1x} \tag{2.9}$$

In matrix form, the above equation can be written as

$$\hat{u} = \begin{bmatrix} N_1 & N_2 \end{bmatrix} \begin{Bmatrix} \hat{d}_{1x} \\ \hat{d}_{2x} \end{Bmatrix} \tag{2.10}$$

with the shape functions:

$$N_1 = 1 - \frac{\hat{x}}{L} \tag{2.11}$$

$$N_2 = \frac{\hat{x}}{L} \tag{2.12}$$

The strain–displacement is

$$\epsilon_x = \frac{d\hat{u}}{d\hat{x}} = \frac{\hat{d}_{2x} - \hat{d}_{1x}}{L}$$

The stiffness matrix is derived as follows

$$T = A\sigma_x \tag{2.13}$$

or

$$T = AE\left(\frac{\hat{d}_{2x} - \hat{d}_{1x}}{L} \right)$$

Also, the nodal force at node number 1 should have a negative sign, as follows:

$$\hat{f}_{1x} = -T \tag{2.14}$$

$$\hat{f}_{1x} = \frac{AE}{L(\hat{d}_{1x} - \hat{d}_{2x})} \tag{2.15}$$

On the other hand, the nodal force at node number 2 should have a positive sign, as follows:

$$\hat{f}_{2x} = T \tag{2.16}$$

$$\hat{f}_{2x} = \frac{AE}{L(\hat{d}_{2x} - \hat{d}_{1x})} \tag{2.17}$$

Expressing the nodal forces in the x-direction in a matrix form

$$\begin{Bmatrix} \hat{f}_{1x} \\ \hat{f}_{2x} \end{Bmatrix} = \frac{AE}{L} \begin{bmatrix} 1 & -1 \\ -1 & 1 \end{bmatrix} \begin{Bmatrix} \hat{d}_{1x} \\ \hat{d}_{2x} \end{Bmatrix} \tag{2.18}$$

and similarly for the nodal forces in the y-direction

$$\begin{Bmatrix} \hat{f}_{1y} \\ \hat{f}_{2y} \end{Bmatrix} = \frac{AE}{L} \begin{bmatrix} 1 & -1 \\ -1 & 1 \end{bmatrix} \begin{Bmatrix} \hat{d}_{1y} \\ \hat{d}_{2y} \end{Bmatrix} \tag{2.19}$$

Since $\underline{\hat{f}} = \underline{\hat{k}}\,\underline{\hat{d}}$, the stiffness matrix for a bar element in a local coordinate can be written as

$$\underline{\hat{k}} = \frac{AE}{L} \begin{bmatrix} 1 & -1 \\ -1 & 1 \end{bmatrix} \tag{2.20}$$

The local coordinate system is always chosen to represent an individual element, while the global coordinate system is chosen for the whole structure. In order to relate the global displacement components to a local one, the transformation matrix is used. Figure 2.1 shows the general displacement vector with local and global coordinate systems.

The angle θ is positive when it is measured clockwise from the global to the local x-axis. The transformation matrix is used to relate the local to the global displacement at node 1, as follows

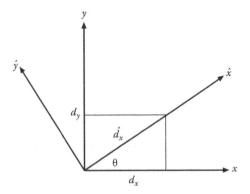

FIGURE 2.1 Relationship between the local and global displacements.

$$\begin{Bmatrix} \hat{d}_{1x} \\ \hat{d}_{1y} \end{Bmatrix} = \begin{bmatrix} \cos\theta & \sin\theta \\ -\sin\theta & \cos\theta \end{bmatrix} \begin{Bmatrix} d_{1x} \\ d_{1y} \end{Bmatrix} \qquad (2.21)$$

and for node 2

$$\begin{Bmatrix} \hat{d}_{2x} \\ \hat{d}_{2y} \end{Bmatrix} = \begin{bmatrix} \cos\theta & \sin\theta \\ -\sin\theta & \cos\theta \end{bmatrix} \begin{Bmatrix} d_{2x} \\ d_{2y} \end{Bmatrix} \qquad (2.22)$$

The global element nodal forces vector is related to the global displacement vector using the global stiffness matrix, as follows

$$\begin{Bmatrix} \hat{f}_{1x} \\ \hat{f}_{1y} \\ \hat{f}_{2x} \\ \hat{f}_{2y} \end{Bmatrix} = \underline{k} \begin{Bmatrix} \hat{d}_{1x} \\ \hat{d}_{1y} \\ \hat{d}_{2x} \\ \hat{d}_{2y} \end{Bmatrix} \qquad (2.23)$$

where

$$\underline{k} = \frac{AE}{L} \begin{bmatrix} 1 & 0 & -1 & 0 \\ 0 & 1 & 0 & -1 \\ -1 & 0 & 1 & 0 \\ 0 & 1 & 0 & -1 \end{bmatrix} \qquad (2.24)$$

From Figure 2.2, the displacement for the nodes 1 and 2 in the local coordinate system can be obtained from its global one using the transformation matrix

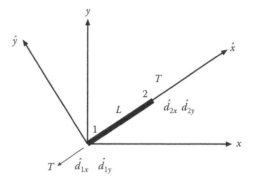

FIGURE 2.2 Bar element subjected to tensile force, and the displacements at the nodes.

$$\hat{d}_{1x} = d_{1x} \cos\theta + d_{1y} \sin\theta \qquad (2.25)$$

$$\hat{d}_{2x} = d_{2x} \cos\theta + d_{2y} \sin\theta \qquad (2.26)$$

and in a matrix form

$$\left\{ \begin{matrix} \hat{d}_x \\ \hat{d}_y \end{matrix} \right\} = \begin{bmatrix} \cos\theta & \sin\theta & 0 & 0 \\ 0 & 0 & \cos\theta & \sin\theta \end{bmatrix} \left\{ \begin{matrix} d_{1x} \\ d_{1y} \\ d_{2x} \\ d_{2y} \end{matrix} \right\} \qquad (2.27)$$

and in a compact form

$$\underline{\hat{d}} = \underline{T}\,\underline{d} \qquad (2.28)$$

At each node, the displacement transformation can be written as

$$\left\{ \begin{matrix} \hat{d}_{1x} \\ \hat{d}_{1y} \\ \hat{d}_{2x} \\ \hat{d}_{2y} \end{matrix} \right\} = \begin{bmatrix} \cos\theta & \sin\theta & 0 & 0 \\ -\sin\theta & \cos\theta & 0 & 0 \\ 0 & 0 & \cos\theta & \sin\theta \\ 0 & 0 & -\sin\theta & \cos\theta \end{bmatrix} \left\{ \begin{matrix} d_{1x} \\ d_{1y} \\ d_{2x} \\ d_{2y} \end{matrix} \right\} \qquad (2.29)$$

Similarly, the nodal force is transformed from the global to the local coordinate system in a similar manner:

$$\begin{Bmatrix} \hat{f}_x \\ \hat{f}_y \end{Bmatrix} = \begin{bmatrix} \cos\theta & \sin\theta & 0 & 0 \\ 0 & 0 & \cos\theta & \sin\theta \end{bmatrix} \begin{Bmatrix} f_{1x} \\ f_{1y} \\ f_{2x} \\ f_{2y} \end{Bmatrix} \tag{2.30}$$

In a compact form:

$$\hat{\underline{f}} = \underline{T}\,\underline{f} \tag{2.31}$$

The force transformation can be written as

$$\begin{Bmatrix} \hat{f}_{1x} \\ \hat{f}_{1y} \\ \hat{f}_{2x} \\ \hat{f}_{2y} \end{Bmatrix} = \begin{bmatrix} \cos\theta & \sin\theta & 0 & 0 \\ -\sin\theta & \cos\theta & 0 & 0 \\ 0 & 0 & \cos\theta & \sin\theta \\ 0 & 0 & -\sin\theta & \cos\theta \end{bmatrix} \begin{Bmatrix} f_{1x} \\ f_{1y} \\ f_{2x} \\ f_{2y} \end{Bmatrix} \tag{2.32}$$

Substituting $\hat{\underline{d}} = \underline{T}\,\underline{d}$ and $\hat{\underline{f}} = \underline{T}\,\underline{f}$ into $\hat{\underline{f}} = \hat{\underline{k}}\,\hat{\underline{d}}$, we obtain

$$\underline{T}\,\underline{f} = \hat{\underline{k}}\underline{T}\,\underline{d} \tag{2.33}$$

By multiplying both sides of the above equation by $\underline{T}^{-1}$, we have

$$\underline{f} = \underline{T}^{-1}\hat{\underline{k}}\underline{T}\,\underline{d} \tag{2.34}$$

Since the transformation matrix is orthogonal, the transpose of $\underline{T}$ is equal to the inverse of $\underline{T}$. Hence

$$\underline{f} = \underline{T}^{\mathrm{T}}\hat{\underline{k}}\underline{T}\,\underline{d} \tag{2.35}$$

Since $\underline{f} = \underline{k}\,\underline{d}$, the global stiffness matrix for an element can be written as

$$\underline{k} = \underline{T}^{\mathrm{T}}\hat{\underline{k}}\underline{T} \tag{2.36}$$

Expanding the stiffness matrix, we obtain

$$k = \frac{AE}{L}\begin{bmatrix} \cos\theta\cos\theta & \cos\theta\sin\theta & -\cos\theta\cos\theta & -\cos\theta\sin\theta \\ \cos\theta\sin\theta & \sin\theta\sin\theta & -\sin\theta\cos\theta & -\sin\theta\sin\theta \\ -\cos\theta\cos\theta & -\sin\theta\cos\theta & \cos\theta\cos\theta & \sin\theta\cos\theta \\ -\cos\theta\sin\theta & -\sin\theta\sin\theta & \sin\theta\cos\theta & \sin\theta\sin\theta \end{bmatrix} \tag{2.37}$$

Assembling the global stiffness and force matrices using the direct stiffness method to obtain

$$\underline{K} = \sum_{e=1}^{N} \underline{k}^{(e)} \tag{2.38}$$

and

$$\underline{F} = \sum_{e=1}^{N} \underline{f}^{(e)} \tag{2.39}$$

The global stiffness matrix is related to the global nodal force matrix and the global displacement matrix for the whole structure by

$$\underline{F} = \underline{K}\,\underline{d} \tag{2.40}$$

Example 2.1:   Solving a plane bar-truss problem

For the plane bar-truss structure shown in Figure 2.3, determine the horizontal and vertical displacements at node number 1. Horizontal and vertical forces are applied at node 1. Given $E = 210\,\text{GPa}$ and $A = 4.0 \times 10^{-4}\,\text{m}^2$.

Step 1:   Constructing the stiffness matrix for each element.
The stiffness matrix for element 1, between nodes 1 and 2 with $\theta = 90°$, is calculated using Equation 2.37:

$$k^1 = \frac{4 \times 10^{-4}(210 \times 10^9)}{3} \begin{bmatrix} 0 & 0 & 0 & 0 \\ 0 & 1 & 0 & -1 \\ 0 & 0 & 0 & 0 \\ 0 & -1 & 0 & 1 \end{bmatrix}$$

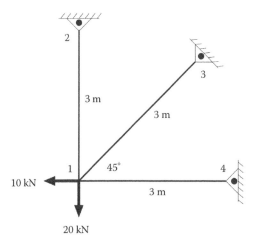

FIGURE 2.3   Bar-truss structure.

The stiffness matrix for element 2, between nodes 1 and 3 with $\theta = 45°$, is calculated using Equation 2.37:

$$k^2 = \frac{4 \times 10^{-4}(210 \times 10^9)}{3} \begin{bmatrix} 0.5 & 0.5 & -0.5 & -0.5 \\ 0.5 & 0.5 & -0.5 & -0.5 \\ -0.5 & -0.5 & 0.5 & 0.5 \\ -0.5 & -0.5 & 0.5 & 0.5 \end{bmatrix}$$

The stiffness matrix for element 3, between nodes 1 and 4 with $\theta = 0°$, is calculated using Equation 2.37:

$$k^3 = \frac{4 \times 10^{-4}(210 \times 10^9)}{3} \begin{bmatrix} 1 & 0 & -1 & 0 \\ 0 & 0 & 0 & 0 \\ -1 & 0 & 1 & 0 \\ 0 & 0 & 0 & 0 \end{bmatrix}$$

Step 2: Assembling the elements' stiffness matrices to form a global stiffness matrix using Equation 2.38:

$$\underline{K} = \frac{4 \times 10^{-4}(210 \times 10^9)}{3} \begin{bmatrix} 1.5 & 0.5 & 0 & 0 & -0.5 & -0.5 & -1 & 0 \\ 0.5 & 1.5 & 0 & -1 & -0.5 & -0.5 & 0 & 0 \\ 0 & 0 & 0 & 0 & 0 & 0 & 0 & 0 \\ 0 & -1 & 0 & 1 & 0 & 0 & 0 & 0 \\ -0.5 & -0.5 & 0 & 0 & 0.5 & 0.5 & 0 & 0 \\ -0.5 & -0.5 & 0 & 0 & 0.5 & 0.5 & 0 & 0 \\ -1 & 0 & 0 & 0 & 0 & 0 & 1 & 0 \\ 0 & 0 & 0 & 0 & 0 & 0 & 0 & 0 \end{bmatrix}$$

Step 3: Applying the boundary conditions to the global stiffness matrix using Equation 2.40:

$$\begin{Bmatrix} f_{1x} = -10 \\ f_{1y} = -20 \\ f_{2x} \\ f_{2y} \\ f_{3x} \\ f_{3y} \\ f_{4x} \\ f_{4y} \end{Bmatrix} = \overline{K} \begin{Bmatrix} d_{1x} \\ d_{1y} \\ d_{2x} = 0 \\ d_{2y} = 0 \\ d_{3x} = 0 \\ d_{3y} = 0 \\ d_{4x} = 0 \\ d_{4y} = 0 \end{Bmatrix}$$

Step 4: Eliminating rows and columns from the global stiffness matrix:

$$\begin{Bmatrix} f_{1x} = -10 \\ f_{1y} = -20 \end{Bmatrix} = \frac{4 \times 10^{-4}(210 \times 10^9)}{3} \begin{bmatrix} 1.5 & 0.5 \\ 0.5 & 1.5 \end{bmatrix} \begin{Bmatrix} d_{1x} \\ d_{1y} \end{Bmatrix}$$

Step 5: *Solving for d_{1x} and d_{1y}:*

$$d_{1x} = -0.0893 \times 10^{-3} \, m$$

$$d_{1y} = -0.4464 \times 10^{-3} \, m$$

2.2   ANALYZING A BAR–TRUSS STRUCTURE

For the plane bar-truss structure shown in Figure 2.4, determine the horizontal and vertical displacements at node number 1 using the ANSYS, and the reactions at the supports. The geometry is similar to the previous example. Horizontal and vertical forces are applied at node number 1, as shown. Given $E = 210\,GPa$ and $A = 4.0 \times 10^{-4} \, m^2$.

Double click on the ANSYS icon

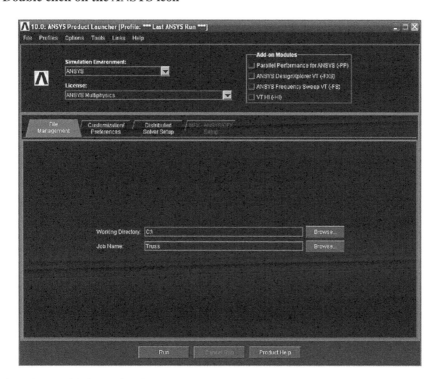

A select Multiphysics in License

B change the Working Directory to C:\

C change the initial Job Name to Truss

Run

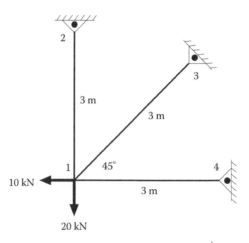

FIGURE 2.4   A plane bar-truss structure.

Main Menu > Preferences

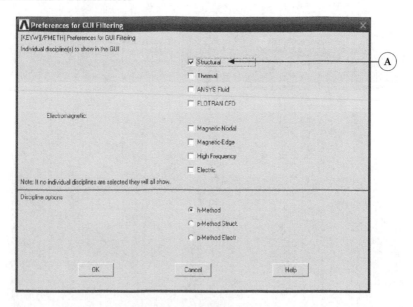

A select the Structural

| OK |

This example is limited to structural analysis. Hence, select Structural. The selected link element will not support pressure on the elements or moment on the nodes and the 2D spar type is limited to two-dimensional analyses.

Main Menu > Preprocessor > Element type > Add/Edit/Delete

Add...

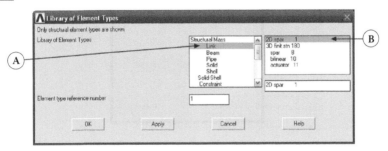

A select Link

B select 2D spar

OK

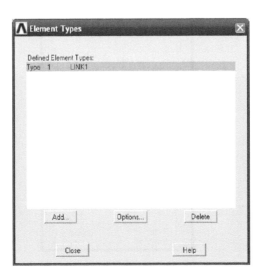

Close

The cross-section area of the bar is required for the analysis. The initial strain is by default zero.

Main Menu > Preprocessor > Real Constants > Add/Edit/Delete

Add...

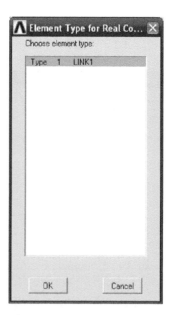

OK

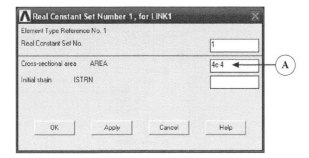

A type 4e-4 in Cross-sectional area

OK

Close

Material properties of the bars are specified in the following steps. The bars are elastic and independent of direction, and are isotropic. Only the modulus of elasticity is required, and the zero Poisson's ratio, or any value to avoid an error message from ANSYS.

Main Menu > Preprocessor > Material Props > Material Models

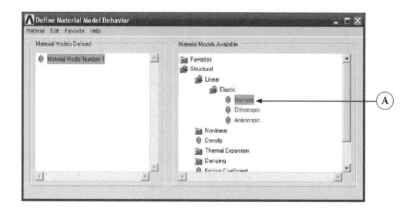

A Double click on Structure > Linear > Elastic > Isotropic

The following windows will show up

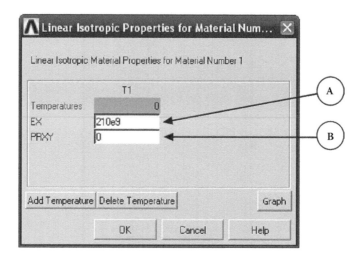

A type 210e9 in EX

B type 0 in the PRXY

[OK]

Close the material model behavior window

The modeling session is started here. First, four nodes are created, followed by creating the elements. The x- and y-coordinate of each node are identified for the ANSYS with its number.

Main Menu > Preprocessor > Modeling > Create > Nodes > In Active CS

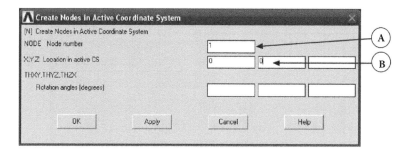

A type 1 in the Node number

B type 0 and 0 in the X, Y, Z Location in active CS

Apply

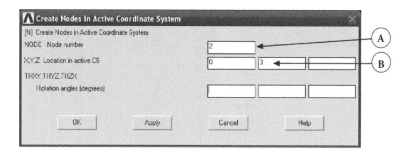

A type 2 in the Node number

B type 0 and 3 in the X, Y, Z Location in active CS

Apply

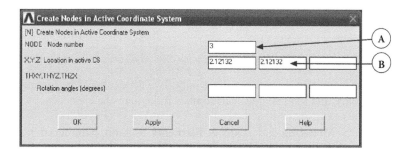

A type 3 in the Node number

B type 2.12132 and 2.12132 in the X, Y, Z Location in active CS

Apply

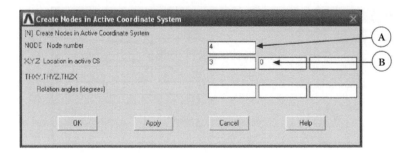

A type 4 in the Node number

B type 3 and 0 in the X, Y, Z Location in active CS

OK

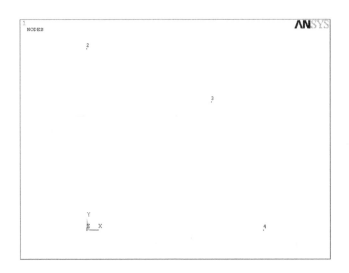

ANSYS graphics shows the created nodes

Creating elements is done using the mouse only. Two nodes should be connected by one element only, and a single node can connect more than one elements.

Main Menu > Modeling > Create > Elements > Auto Numbered > Thru Nodes

Click on node 1 then 2, Apply

Click on node 1 then 3, Apply

Click on node 1 then 4, OK

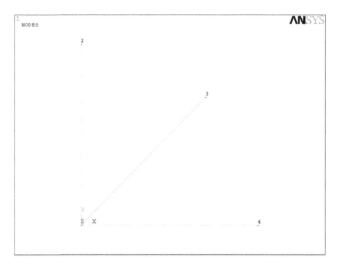

ANSYS graphics shows the elements

The solution task starts here. Nodal forces and displacements are applied. Starting with forces or displacements will not affect the solution. A zero displacement at a node means that the node is fixed.

Main Menu > Solution > Define Load > Apply > Structural > Displacement > On Nodes

Click on nodes 2, 3, and 4

OK

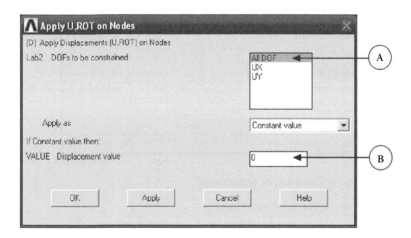

A select All DOF

B type 0 in the Displacement value

OK

Main Menu > Solution > Define Load > Apply > Structural > Force/Moment > On Nodes

Click on Node 1

OK

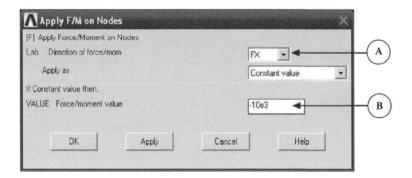

A select FX in the Direction of force/mom

B type -10e3 in the Force/moment value

Apply

Click again on node number 1

OK

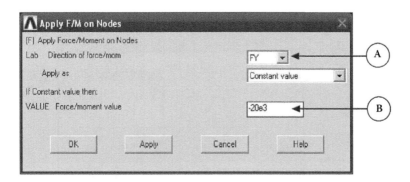

A select FY in the Direction of force/mom

B type -20e3 in the Force/moment value

OK

The negative FX means that the force at node 1 is in the negative *x*-direction, and negative FY means that the force at node 1 is in the negative *y*-direction. Clicking Apply button will not close the window, but allows for additional inputs. Clicking OK will close the window. The ANSYS graphics will show the applied forces and their direction by a red arrow. Reapplying the force at the node 1 will automatically delete the force input and apply the new input.

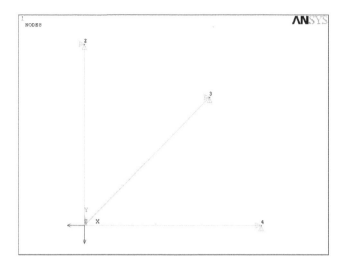

ANSYS graphics shows the forces with direction and the displacement on nodes

The final step is to initiate the solution. ANSYS will assemble the stiffness matrix, apply the boundary conditions, and solve the stiffness matrix.

Main Menu > Solution > Solve > Current LS

OK

Close

Results can be viewed and listed in the general postprocessor task. Inspecting the deformation of the bar structure will help to identify if any boundary condition was wrongly applied.

Main Menu > General Postproc > Plot Results > Deformed Shape

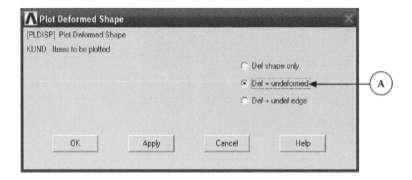

A select Def + undeformed

 OK

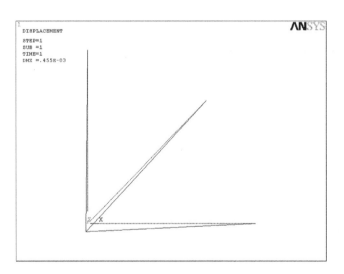

ANSYS graphics shows the bar before and after deformation

From the above figure, the result is as expected. Nodes 2, 3, and 4 are fixed, while node 1 is moved in the direction of the applied forces. The following steps are devoted to list the nodal results.

Main Menu > General Postproc > List Results > Nodal Solution

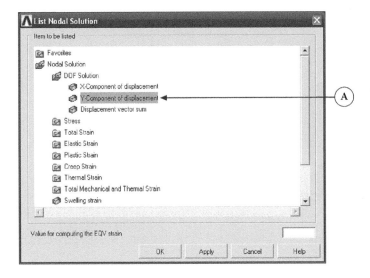

A select Nodal Solution > DOF Solution > Y-Component of displacement

OK

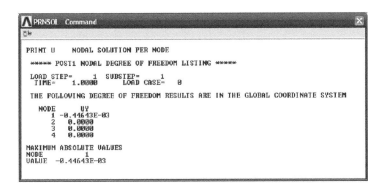

A list of nodal displacement in the *y*-direction is shown. In addition, the maximum nodal displacement is shown at the end. The displacement in the *x*-direction can be determined by selecting X-Component. Notice the results are identical to the results of the previous example.

Main Menu > General Postproc > List Results > Reaction Solution

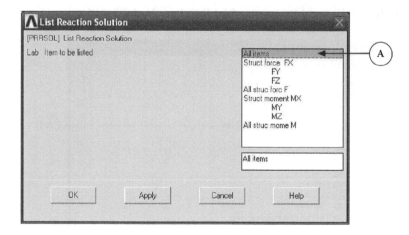

A select All items

OK

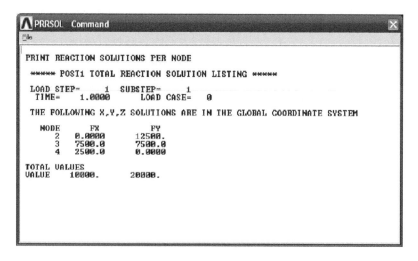

From the reaction results at the supports, the forces are balanced in the x- and y-directions. Results for node number 1 is not shown because force is applied on it.

2.3 DEVELOPMENT OF HORIZONTAL BEAM ELEMENTS

A beam is defined as a long and slender structural member that can be subjected to transverse loading. The bar elements theory that is discussed in Section 2.1 can sustain transverse loading only. Consider the beam element shown in Figure 2.5. The beam has a length L, and the local displacement and rotation at nodes 1 and 2 are $(\hat{d}_{1y}, \hat{\varnothing}_1)$ and $(\hat{d}_{2y}, \hat{\varnothing}_2)$, respectively. The local nodal forces and bending moments

FIGURE 2.5 Beam element subjected to forces and moments.

at nodes 1 and 2 are $(\hat{f}_{1y}, \hat{m}_1)$ and $(\hat{f}_{2y}, \hat{m}_2)$, respectively. It is assumed that the beam has an initial length L, constant modulus of elasticity E, and uniform moment of inertia I.

All moments and rotations are positive if their direction is counterclockwise and negative if their direction is clockwise. The horizontal forces are positive if the direction is positive in the local x-direction and negative if the direction is negative in the local x-direction. The vertical forces are positive if the direction is positive in the local y-direction and negative if the direction is negative in the local y-direction. Consider a differential beam element subjected to a distributed pressure loading $w(\hat{x})$, as shown in Figure 2.6.

Applying the force balance in the local y-direction, we obtain

$$V - \left(V + dV\right) - w\left(\hat{x}\right)d\hat{x} = 0 \tag{2.41}$$

or

$$w(\hat{x}) = -\frac{dV}{d\hat{x}} \tag{2.42}$$

Applying the moments balance at node 2, we obtain

$$-V\,dx + dM + w(\hat{x})d\hat{x}\left(\frac{d\hat{x}}{2}\right) = 0 \tag{2.43}$$

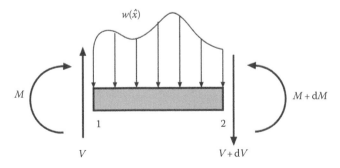

FIGURE 2.6 A differential beam element.

or

$$V = -\frac{dM}{d\hat{x}} \qquad (2.44)$$

The curvature of the beam can be related to the applied moment. The radius of the deflective curve ρ is varied along the beam, while Ø is the rotation, and the function $\hat{v}(\hat{x})$ is the transverse displacement in the $\hat{y}$-direction. Figure 2.7 shows the deflected beam.

The radius of the deflection can be expressed as follows:

$$\frac{1}{\rho} = \frac{M}{EI} \qquad (2.45)$$

The rotation of the deflection can be expressed as

$$\hat{\varnothing} = \frac{d\hat{v}(\hat{x})}{d\hat{x}} \qquad (2.46)$$

The axial strain is related to the axial displacement using the following relationship:

$$\epsilon_x = \frac{d\hat{u}}{d\hat{x}} \qquad (2.47)$$

On the other hand, the axial displacement is related to the transverse displacement by

$$\hat{u} = -\frac{\hat{y}(d\hat{v})}{d\hat{x}} \qquad (2.48)$$

Substituting Equation 2.48 into Equation 2.47, yields

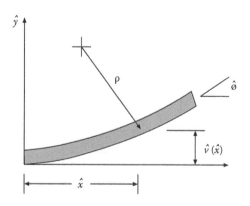

FIGURE 2.7   Radius and rotation at $\hat{v}(\hat{x})$.

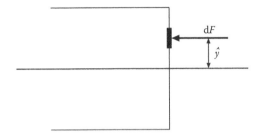

FIGURE 2.8 Beam cross section.

$$\epsilon_x = -\frac{\hat{y}(d^2\hat{v})}{d\hat{x}^2}$$ (2.49)

For the cross-section area of the beam shown in Figure 2.8, $dF = \sigma_x\,dA$, the differential moment along the center of the beam is equal to

$$dM = \sigma_x\hat{y}\,dA$$ (2.50)

Using Hooke's equation and Equation 2.49, the differential moment becomes

$$dM = E\epsilon_x y\,dA = -E\hat{y}\frac{\hat{y}(d^2\hat{v})}{d\hat{x}^2}\,dA$$ (2.51)

Integrating Equation 2.51

$$M = E\frac{d^2\hat{v}}{d\hat{x}^2}\int\hat{y}^2\,dA$$ (2.52)

Since the moment of inertia is equal to $\int y^2\,dA$, Equation 2.52 becomes

$$M = EI\frac{d^2\hat{v}}{d\hat{x}^2}$$ (2.53)

Also, the shearing force can be expressed as

$$V = EI\frac{d^3\hat{v}}{d\hat{x}^3}$$ (2.54)

The variation of curvature of the beam is assumed polynomial of the third order, and the constants in the polynomial are determined using the boundary conditions, as follows:

$$\hat{v} = a_1\hat{x}^3 + a_2\hat{x}^2 + a_3\hat{x} + a_4 \tag{2.55}$$

Applying the boundary conditions

$$\hat{v}(0) = \hat{d}_{1y} : a_4 = \hat{d}_{1y}$$

$$\hat{v}(L) = \hat{d}_{2y} : a_1L^3 + a_2L^2 + a_3L + a_4 = \hat{d}_{2y}$$

$$\hat{\varnothing}(0) = \hat{\varnothing}_1 : a_3 = \hat{\varnothing}_1$$

$$\hat{\varnothing}(L) = \hat{\varnothing}_2 : 3a_1L^2 + 2a_2L + a_3 = \hat{\varnothing}_2$$

Solving the above equations for a_1, a_2, a_3, and a_4 and substituting the solution into Equation 2.55, we have

$$\hat{v}(\hat{x}) = \left[\frac{2}{L^3}(\hat{d}_{1y} - \hat{d}_{2y}) + \frac{1}{L^2}(\hat{\varnothing}_1 + \hat{\varnothing}_2)\right]\hat{x}^3 + \left[-\frac{3}{L^2}(\hat{d}_{1y} - \hat{d}_{2y}) - \frac{1}{L}(2\hat{\varnothing}_1 - \hat{\varnothing}_2)\right]\hat{x}^2$$
$$+ \hat{\varnothing}_1\hat{x} + \hat{d}_{1y} \tag{2.56}$$

Finally, the element stiffness matrix is derived using the equilibrium approach. The nodal shear force and bending moment are calculated using Equations 2.53 and 2.54, respectively, as follows:

$$\hat{f}_{1y} = \hat{V} = EI\frac{d^3\hat{v}(0)}{d\hat{x}^3} = \frac{EI}{L^3}\left(12\hat{d}_{1y} + 6L\hat{\varnothing}_1 - 12\hat{d}_{2y} + 6L\hat{\varnothing}_2\right) \tag{2.57}$$

$$\hat{m}_1 = -\hat{m} = -EI\frac{d^2\hat{v}(0)}{d\hat{x}^2} = \frac{EI}{L^3}\left(6\hat{d}_{1y} + 4L^2\hat{\varnothing}_1 - 6\hat{d}_{2y} + 2L^2\hat{\varnothing}_2\right) \tag{2.58}$$

$$\hat{f}_{2y} = -\hat{V} = -EI\frac{d^3\hat{v}(L)}{d\hat{x}^3} = \frac{EI}{L^3}\left(-12\hat{d}_{1y} - 6L\hat{\varnothing}_1 + 12\hat{d}_{2y} - 6L\hat{\varnothing}_2\right) \tag{2.59}$$

$$\hat{m}_2 = \hat{m} = EI\frac{d^2\hat{v}(L)}{d\hat{x}^2} = \frac{EI}{L^3}\left(6\hat{d}_{1y} + 2L^2\hat{\varnothing}_1 - 6\hat{d}_{2y} + 4L^2\hat{\varnothing}_2\right) \tag{2.60}$$

The above results can be expressed in a matrix form, $\underline{F} = \underline{K}\,\underline{d}$; the matrix expression will have the following form

$$\begin{Bmatrix} \hat{f}_{1y} \\ \hat{m}_1 \\ \hat{f}_{2y} \\ \hat{m}_2 \end{Bmatrix} = \frac{EI}{L^3} \begin{bmatrix} 12 & 6L & -12 & 6L \\ 6L & 4L^2 & -6L & 2L^2 \\ -12 & -6L & 12 & -6L \\ 6L & 2L^2 & -6L & 4L^2 \end{bmatrix} \begin{Bmatrix} \hat{d}_{1y} \\ \hat{\varnothing}_1 \\ \hat{d}_{2y} \\ \hat{\varnothing}_2 \end{Bmatrix} \quad (2.61)$$

where the local stiffness matrix is

$$\underline{\hat{k}} = \frac{EI}{L^3} \begin{bmatrix} 12 & 6L & -12 & 6L \\ 6L & 4L^2 & -6L & 2L^2 \\ -12 & -6L & 12 & -6L \\ 6L & 2L^2 & -6L & 4L^2 \end{bmatrix} \quad (2.62)$$

Example 2.2: Solving horizontal beam structure

For the horizontal beam shown in Figure 2.9, determine the displacement and slopes at the node number 2. Force and moment are applied at node 2, as shown in Figure 2.9. Given $E = 210\,GPa$ and $I = 4 \times 10^{-4}\,m^4$.

First, the stiffness matrix for each of the two elements is created using Equation 2.62:

$$k^1 = \frac{210 \times 10^9 (4 \times 10^4)}{3^3} \begin{bmatrix} 12 & 18 & -12 & 18 \\ 18 & 36 & -18 & 18 \\ -12 & -18 & 12 & -18 \\ 18 & 18 & -18 & 36 \end{bmatrix}$$

$$k^2 = \frac{210 \times 10^9 (4 \times 10^4)}{3^3} \begin{bmatrix} 12 & 18 & -12 & 18 \\ 18 & 36 & -18 & 18 \\ -12 & -18 & 12 & -18 \\ 18 & 18 & -18 & 36 \end{bmatrix}$$

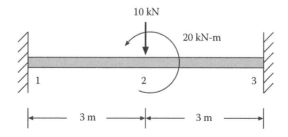

FIGURE 2.9 Beam structure for Example 2.2.

Assembling the above two matrices using Equations 2.38 through 2.40:

$$\begin{Bmatrix} \hat{f}_{1y} \\ \hat{m}_1 \\ \hat{f}_{2y} = -10\times10^3 \\ \hat{m}_2 = 20\times10^3 \\ \hat{f}_{3y} \\ \hat{m}_3 \end{Bmatrix} = \frac{210\times10^9(4\times10^4)}{3^3} \begin{bmatrix} 12 & 18 & -12 & 18 & 0 & 0 \\ 18 & 36 & -18 & 18 & 0 & 0 \\ -12 & -18 & 24 & 0 & -12 & 18 \\ 18 & 18 & 0 & 72 & -18 & 18 \\ 0 & 0 & -12 & -18 & 12 & -18 \\ 0 & 0 & 18 & 18 & -18 & 36 \end{bmatrix} \begin{Bmatrix} \hat{d}_{1y} = 0 \\ \hat{\emptyset}_1 = 0 \\ \hat{d}_{2y} \\ \hat{\emptyset}_2 \\ \hat{d}_{3y} = 0 \\ \hat{\emptyset}_3 = 0 \end{Bmatrix}$$

Finally, solving the for $\hat{d}_{2y}$ and $\hat{\emptyset}_2$, the results are

$$\hat{d}_{2y} = -1.34\times10^{-4}\text{ m} \quad \text{and} \quad \hat{\emptyset}_2 = 8.928\times10^{-5}\text{ rad}$$

2.4 ANALYZING A HORIZONTAL BEAM STRUCTURE

For the horizontal beam in Figure 2.10, use ANSYS to determine the displacement and the slopes at the node numbers 1 and 2, and the reactions at the support.
Force, distributed pressure, and moment are applied as shown in the figure. Given
$E = 210\text{ GPa}$, $A = 2 \times 10^{-3}$ m², and $I_{zz} = 4 \times 10^{-4}$ m.

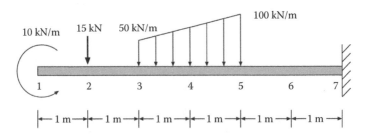

FIGURE 2.10 Beam structure.

Double click on the ANSYS icon

Main Menu > Preferences

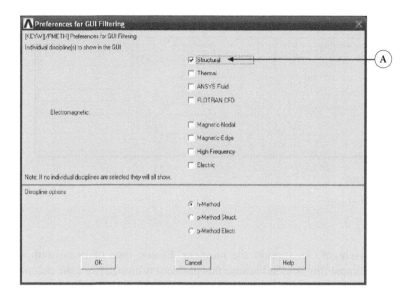

A select the Structural

OK

Main Menu > Preprocessor > Element type > Add/Edit/Delete

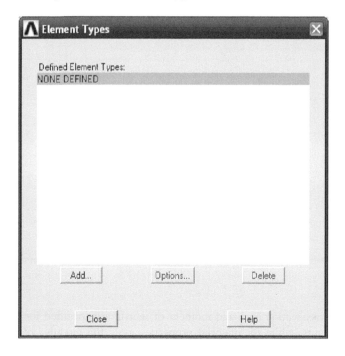

Add...

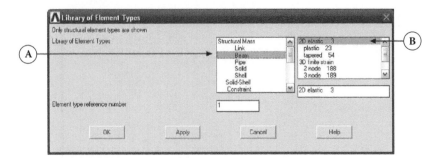

A select Beam

B select 2D elastic

OK

The beam element can support the moment. Elastic type of beam with a two-dimensional capability is used because this problem is limited to elastic deformation in two-dimensional space.

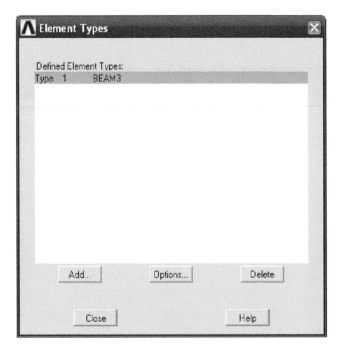

Close

The beam cross-section area and moment of inertia are specified for the ANSYS in real constants, while for material properties, only the modulus of elasticity is required, and any value for the Poisson's ratio is required to avoid an error message from ANSYS.

Main Menu > Preprocessor > Real Constants > Add/Edit/Delete

Add...

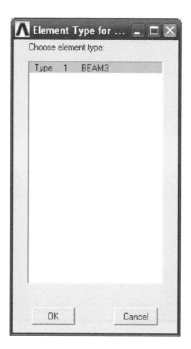

OK

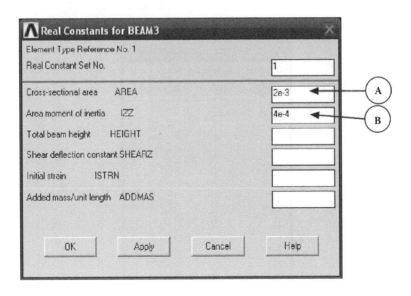

A type 2e-3 in Area

B type 4.0e-4 in Area moment of inertia

OK

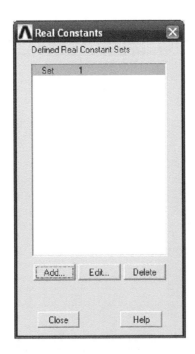

Close

Main Menu > Preprocessor > Material Props > Material Models

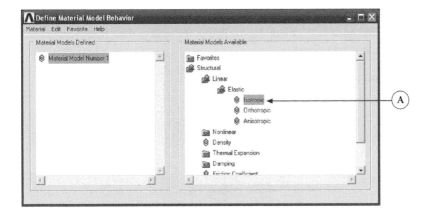

A Double click on Structure > Linear > Elastic > Isotropic

The following window will appear:

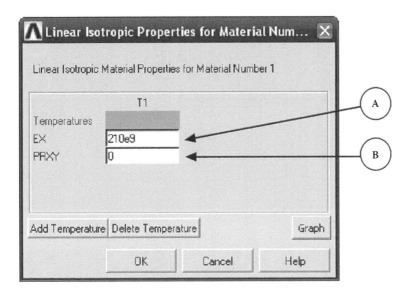

A type 210e9 in EX

B type 0 in the PRXY

Close the material model behavior window

The nodes are created using the work plane. The size of the work plane is six and the space between the grids is one. In the WP setting, the number 0 for minimum means that no grid will be in the negative *x–y* plane. Number 6 in maximum and number 1 in spacing mean that grids will be divided into six-by-six squares and the sides' length of the squares is one. The snap increment is equal to the grid spacing to insure that the node will exactly be located at the grid points. Nodes are created by clicking on the grid points.

Utility Menu > WorkPlane > WP Setting

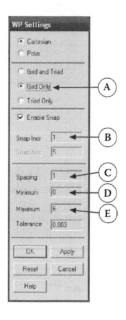

A select Grid only

B type 1 in Snap Incr

C type 1 in Spacing

D type 0 in Minimum

E type 6 in Maximum

OK

Activating the work plane is required to show the grids. The grids can be temporarily disappeared by deactivating the work plane.

ANSYS Utility Menu > WorkPlane > Display Working Plane

ANSYS Utility Menu > PlotCtrls > Pan, Zoom, Rotate

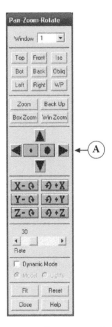

A click on zoom in and zoom out, moving cursor button until the ANSYS graphics shows all grids.

| Close |

ANSYS Main Menu > Preprocessor > Modeling > Create > Nodes > On Working Plane

Using the mouse, click on the ANSYS graphics window at the location of nodes 1–7.

| OK |

ANSYS Utility Menu > WorkPlane > Display Working Plane

Notice that minimum seven nodes are needed to model the beam. If more nodes are used, such as inserting nodes between 1 and 2, the ANSYS results will definitely be more accurate.

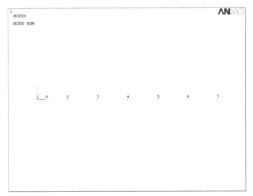

ANSYS graphics shows the nodes with their number

Main Menu > Modeling > Create > Elements > Auto Numbered > Thru Nodes

Click on node 1 then 2, |**Apply**|

Click on node 2 then 3, |**Apply**|

Click on node 3 then 4, |**Apply**|

Click on node 4 then 5, |**Apply**|

Click on node 5 then 6, |**Apply**|

Click on node 6 then 7, |**OK**|

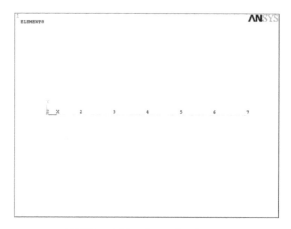

ANSYS graphics shows the elements

Modeling is completed at this point. Next, the boundary conditions are applied starting with the support, then the force and moment, and finally the pressure. This order is not important for solution.

Main Menu > Solution > Define Load > Apply > Structural > Displacement > On Nodes

Click on node number 7

|**OK**|

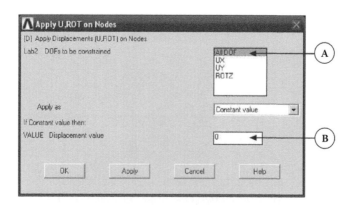

A select All DOF

B type 0 in the Displacement value

**Main Menu > Solution > Define Load > Apply > Structural > Force/Moment >
On Nodes**

Click on node number 2

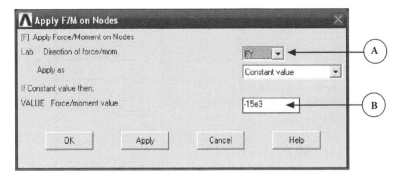

A select FY in the Direction of force/mom

B type -15e3 in the Force/moment value

`OK`

**Main Menu > Solution > Define Load > Apply > Structural > Force/Moment >
On Nodes**

Click on node number 1

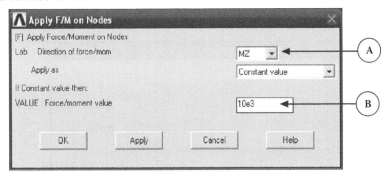

A select MZ in the Direction of force/mom

B type 10e3 in the Force/moment value

`OK`

The positive moment means its direction is counterclockwise. The pressure variation
between nodes 3 and 4 is linear. The value of *I* in the following windows is the pressure

at node 3, and the value of *J* is the pressure at node 4. ANSYS will automatically distribute a linear variation of pressure between the two nodes. The value of *I* should be assigned to the smallest node number in the selected element.

Main Menu > Solution > Define Load > Apply > Structural > Pressure > On beam

Click on the element number 3, which is for nodes 3 and 4

OK

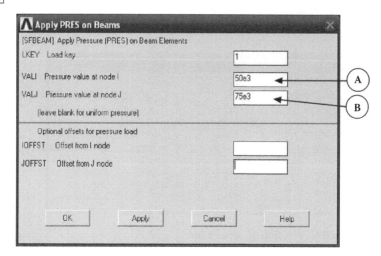

A type 50e3 in Pressure at node I

B type 75e3 in Pressure at node J

Apply

Click on the element number 4, which is between nodes 4 and 5

OK

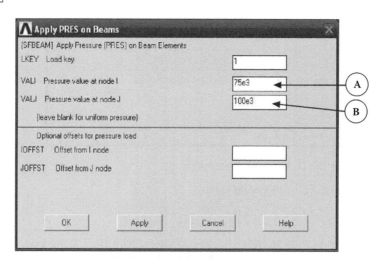

A type 75e3 in Pressure at node I

B type 100e3 in Pressure at node J

OK

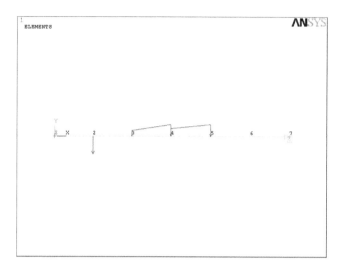

ANSYS graphics shows the pressure, force and moment with direction, and the displacement

The final step is to run the ANSYS solution. ANSYS will assemble the stiffness matrix, apply the boundary conditions, and solve the stiffness matrix.

Main Menu > Solution > Solve > Current LS

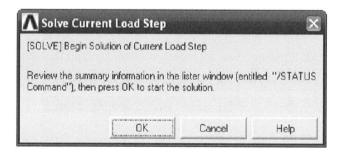

OK

Close

Main Menu > General Postproc > Plot Results > Deformed Shape

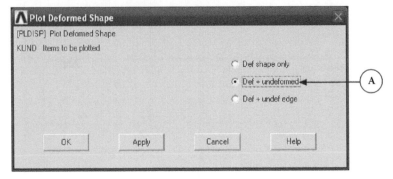

A select Def + undeformed

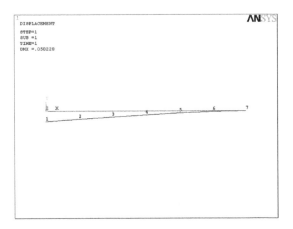

ANSYS graphics shows the beam before and after apply the loads

Main Menu > General Postproc > List Results > Nodal Solution

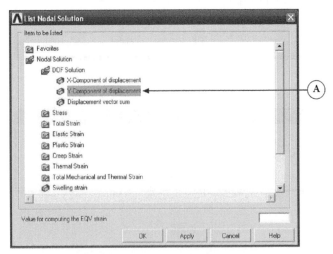

A select Nodal Solution > DOF Solution > Y-Component displacement

OK

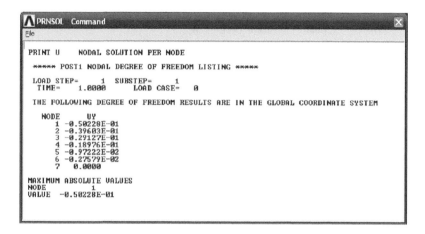

A list of nodal displacements in the *y*-direction is shown. The maximum nodal displacement is shown at the end. Notice that the displacement of the node number 7 is 0 because it is fixed, and node 1 has the highest deflection, as expected. The nodal reactions are displayed in the following steps.

Main Menu > General Postproc > List Results > Reaction Solu

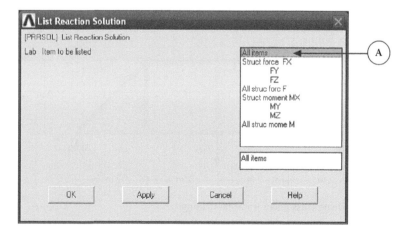

A select All items

OK

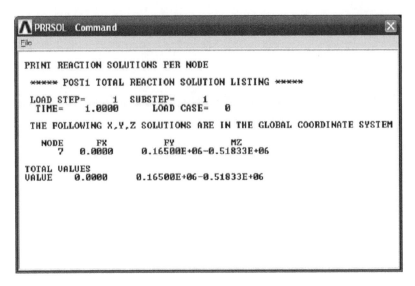

The results are for node 7 only, and the moment reaction is listed.

2.5   BEAM–TRUSS STRUCTURE UNDER A TRANSIENT LOADING

For the beam truss shown in Figure 2.11, a transient force, with a frequency $f = 0.01\,\text{Hz}$ and an amplitude Ap = 25 kN, is oscillating harmonically around a steady load Fs = 25 kN. Use ANSYS to determine the displacement at the point where the force is applied as a function of time. Also, create an animation file for the loading process. The total time for the loading process is 500 s. Given $E = 190\,\text{GPa}$, $A = 2 \times 10^{-2}\,\text{m}^2$, and $I_{zz} = 5 \times 10^{-4}\,\text{m}^4$. Use the following formula to simulate the applied transient force:

$$F(t) = \text{Fs} + \text{Ap}\sin(2\pi f t)$$

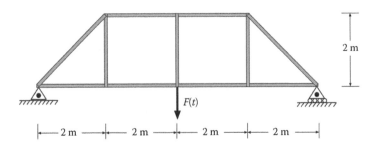

FIGURE 2.11   Beam truss under transient loading.

Double click on the ANSYS icon

Main Menu > Preferences

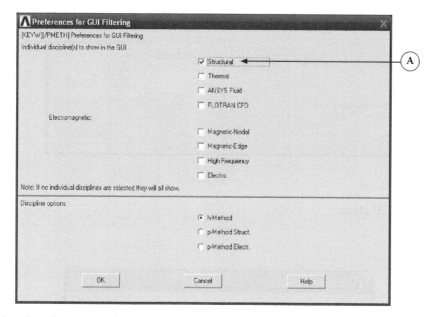

A select the Structural

| OK |

Main Menu > Preprocessor > Element type > Add/Edit/Delete

| Add... |

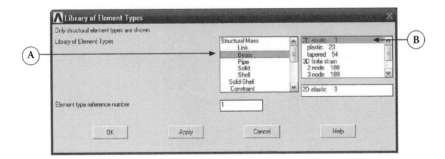

A select Beam

B select 2D elastic

OK

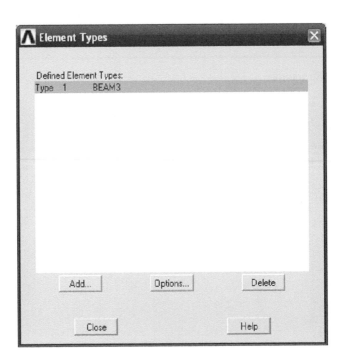

Close

The selected beam element can support the moment. Elastic type of beam with a two-dimensional capability is used because this problem is limited to the elastic deformation in two-dimensional space. The cross-section area of the beam and its moment of inertia are required to solve the problem.

Main Menu > Preprocessor > Real Constants > Add/Edit/Delete

Add...

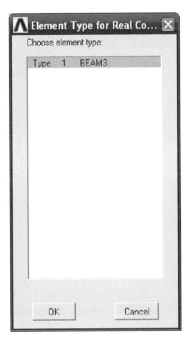

OK

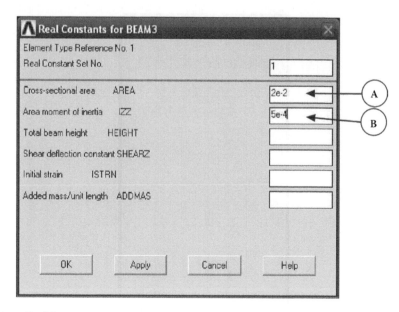

A type 2e-2 in area

B type 5e-4 in Area moment of inertia

OK

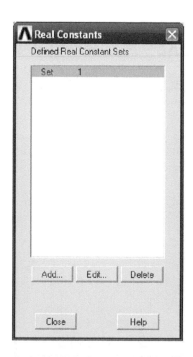

Close

For material properties, only the modulus of elasticity is required, and any value for the Poisson's ratio is required to avoid an error message from the ANSYS.

Main Menu > Preprocessor > Material Props > Material Models

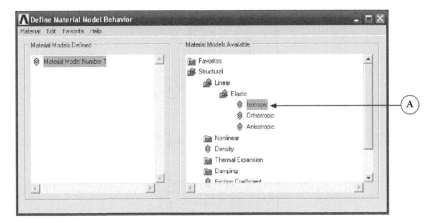

A Double click on Structure > Linear > Elastic > Isotropic

The following window will appear:

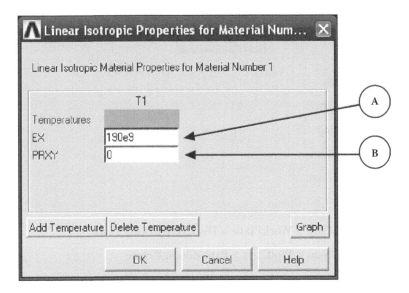

A type 190e9 in EX

B type 0 in the PRXY

Close the material model behavior window

Here, modeling is entirely done using the ANSYS graphics. The nodes and elements are created using the work plane. The size of the work plane is eight and the space between the grids is two.

Utility Menu > WorkPlane > WP Setting

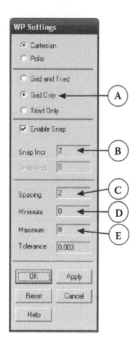

A select Grid Only

B type 2 in Snap Incr

C type 2 in Spacing

D type 0 in Minimum

E type 8 in Maximum

OK

ANSYS Utility Menu > WorkPlane > Display Working Plane

ANSYS Utility Menu > PlotCtrls > Pan, Zoom, Rotate

Click on zoom in and zoom out, moving the cursor button until the ANSYS graphics shows all grids.

Close

ANSYS Main Menu > Preprocessor > Modeling > Create > Nodes > On Working Plane

Using the mouse, click on the ANSYS graphics window at the location of node numbers 1–8.

OK

ANSYS Utility Menu > WorkPlane > Display Working Plane

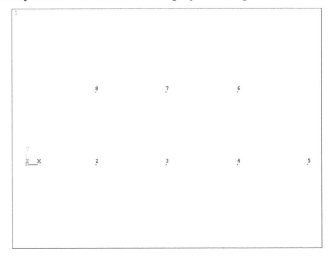

ANSYS graphics shows the nodes with their number

Main Menu > Modeling > Create > Elements > Auto Numbered > Thru Nodes

Click on node 1 then 2, **Apply**

and similarly for all elements, as shown in the figure below.

OK

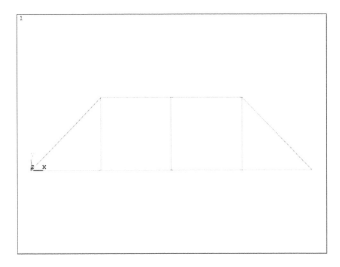

ANSYS graphics shows the elements

This problem is a transient structural analysis. In the new analysis option, the type of the analysis should be changed from static to transient. Then, the solution control inputs will be activated to specify the transient input parameters such as the time duration and time step size.

Main Menu > Solution > Analysis Type > New Analysis

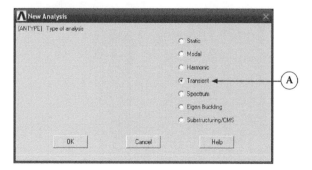

A select Transient

 OK

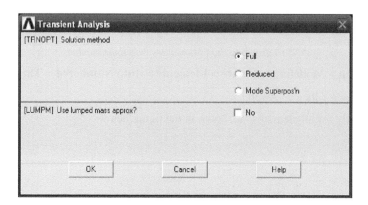

OK

Main Menu > Solution > Analysis Type > Sol'n Control

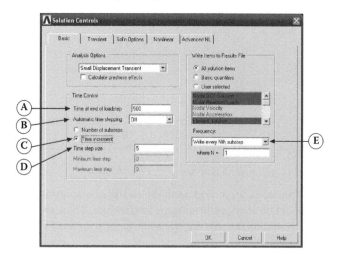

A type 500 in Time at end of loadstep

B turn Off the Automatic time stepping

C select Time increment

D type 5 in Time step size

E select Write every Nth substeps

$\boxed{\text{OK}}$

By selecting the Nth substeps in the frequency, the results at every time step will be stored in the ANSYS result file. Otherwise, only the result at time = 500 s will be stored. The function editor is used to apply transient force formula as a nodal force. This technique is simple and convenient for this problem since an equation for the force is given. Notice that the negative FY means that the force is in the negative y-direction.

Main Menu > Solution > Define Load > Apply > Functions > Define/Edit

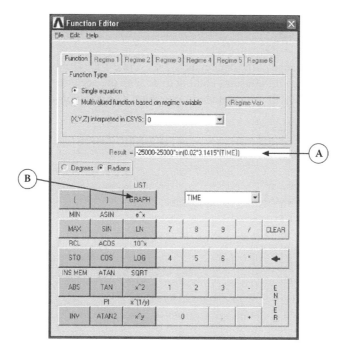

A type the equation: −25000−25000*sin(0.02*3.1415*{TIME})

B click on GRAPH

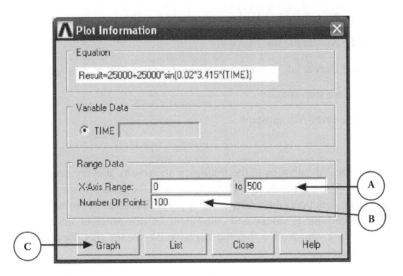

A type 0 and 500 for X-Axis Range

B type 100 in the Number Of Points

C click on Graph

$\boxed{\text{OK}}$

0 and 500 are the range of the data in the x-axis, while 100 is the number of points to be plotted. Number Of Points has nothing to do with the accuracy of the results, and higher number of it will just create a smother plot. The equation should be saved.

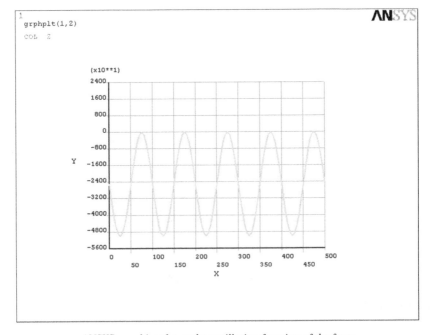

ANSYS graphics shows the oscillating function of the force

In the Equation Editor click on File then Save

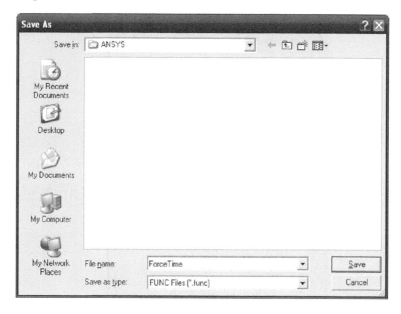

Save the file as ForceTime, and the file name is optional

| Save |

After saving the function, it is required to load it to the ANSYS solution using the read file.

Close the equation editor

Main Menu > Solution > Define Load > Apply > Functions > Read File

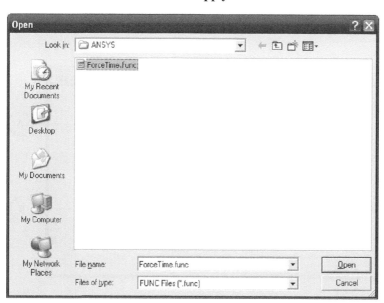

A select ForceTime.func

[**Open**]

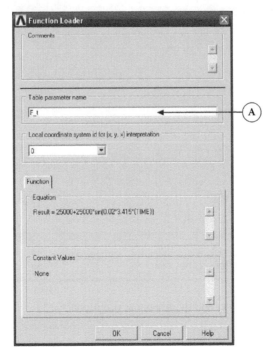

A type F_t in the Table parameter name, and this name is optional.

[OK]

Main Menu > Solution > Define Load > Apply > Structural > Displacement > On Nodes

Click on node number 1

[OK]

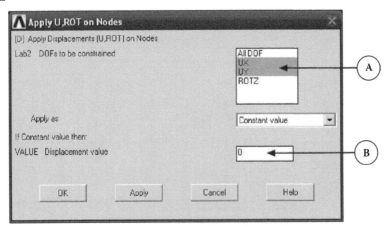

A select UX and UY

B type 0 in the Displacement value

OK

Main Menu > Solution > Define Load > Apply > Structural > Displacement > On Nodes

Click on node number 5

OK

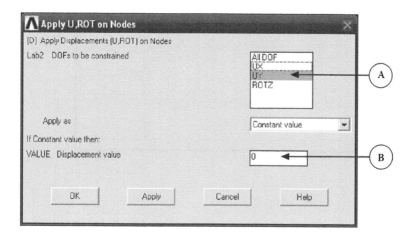

A select UY

B type 0 in the Displacement value

OK

Main Menu > Solution > Define Load > Apply > Structural > Force/Moment > On Nodes

Click on node number 3

OK

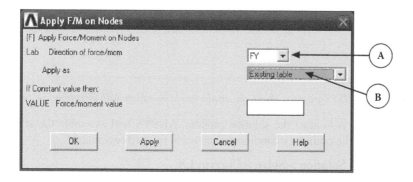

A select FY in the Direction of force/mom

B select Excising table

OK

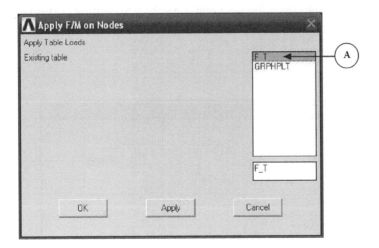

A select F_T

OK

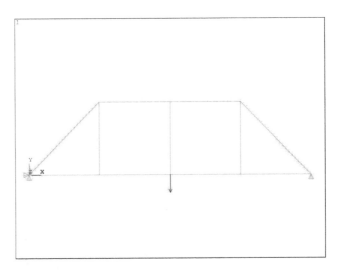

ANSYS graphics shows the displacement and force with direction

The final step is to start the solution process. ANSYS will assemble the stiffness matrix, apply the boundary conditions, and solve the stiffness matrix.

Main Menu > Solution > Solve > Current LS

OK

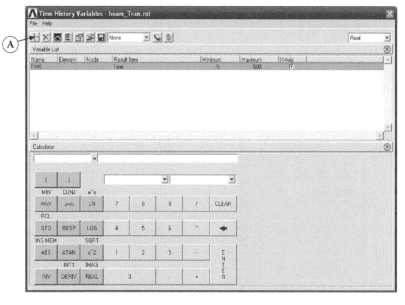

Close

The displacement history at the point where the force is applied is presented. In the time history of the postprocessor, the location of the node is selected first, and then a graph is created in the ANSYS graphics. The displacement history can also be listed.

Main Menu > TimeHist Postpro

A click on the green X button

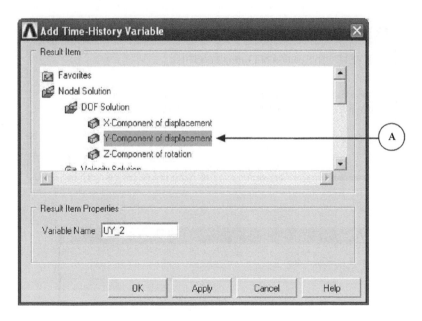

A select Y-Component of displacement

 OK

Click on the node where the force is applied

 OK

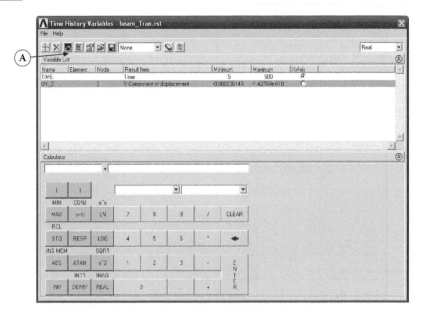

A click on the graph button

 OK

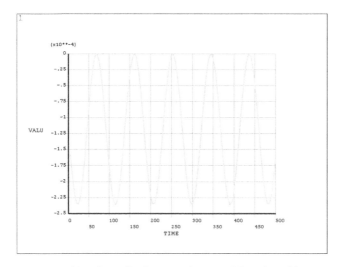

ANSYS graphics shows displacement history of the selected location

Animation of the deformed structure from time = 0 to 500 s can be easily accomplished using animate command in the PlotCtrol options. The number of the frame in the animate over time is the number of pictures in the avi file, while the animation time delay is the display period between two pictures. One hundred animation frames produce a good resolution animation file, and with 0.5 s the animation file duration is 50 s (100 frames × 0.5 delay).

Main Menu > General Postproc

Utility Menu > PlotCtrls > Animate > Over time...

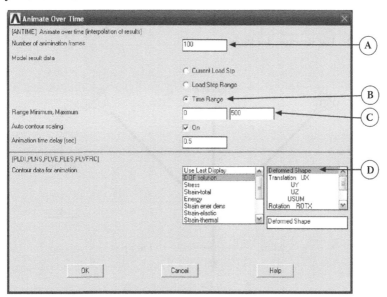

A type 100 in Number of animation frames

B select Time Range

C type 0 and 500 for range Minimum, Maximum

D select Deformed Shape

OK

PROBLEMS

2.1 For the bar structure shown in Figure 2.12, determine the reactions at the supports, and nodal displacements in the y-direction. Let $E = 210\,\text{GPa}$ and $A = 4.0 \times 10^{-4}\,\text{m}^2$. The applied force is $150\,\text{kN}$.

2.2 For the bar structure shown in Figure 2.13, determine the reactions at the supports, and nodal displacements in the y-direction. Let $E = 210\,\text{GPa}$ and $A = 4.0 \times 10^{-4}\,\text{m}^2$.

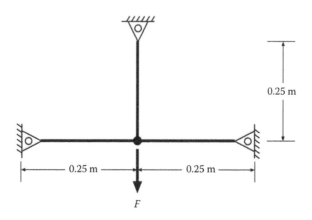

FIGURE 2.12   Bar structure for Problem 2.1.

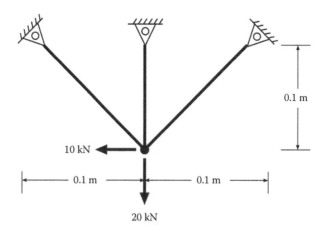

FIGURE 2.13   Bar structure for Problem 2.2.

2.3 Determine the displacement at the points of the applied forces for horizontal beam shown in Figure 2.14, and the reactions at the supports. Given $E = 220\,\text{GPa}$, $A = 2.0 \times 10^{-4}\,\text{m}^2$, and $I_{zz} = 4 \times 10^{-4}\,\text{m}^4$.

2.4 Plot the deflection curve for the beam shown in Figure 2.15, and determine the reactions at the supports. Given $E = 200\,\text{GPa}$, $A = 3.0 \times 10^{-4}\,\text{m}^2$, and $I_{zz} = 4 \times 10^{-4}\,\text{m}^4$.

2.5 For beam-truss structure shown in Figure 2.16, determine the maximum nodal displacement, and identify its location. Also, print out the deformed structure. Given $E = 230\,\text{GPa}$, $A = 3.5 \times 10^{-2}$, and $I_{zz} = 3 \times 10^{-4}\,\text{m}^4$.

2.6 Using ANSYS, obtain the maximum displacement in the y-direction, and the reactions at the supports for the beam truss shown in Figure 2.17. Given $E = 180\,\text{GPa}$, $A = 3.0 \times 10^{-4}\,\text{m}^2$, and $I_{zz} = 3.5 \times 10^{-4}\,\text{m}^4$.

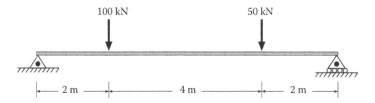

FIGURE 2.14   Horizontal beam structure for Problem 2.3.

FIGURE 2.15   Horizontal beam structure for Problem 2.4.

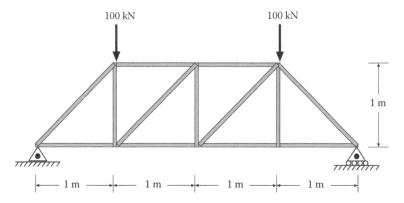

FIGURE 2.16   Beam structure for Problem 2.5.

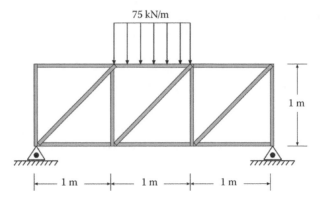

FIGURE 2.17 Beam structure for Problem 2.6.

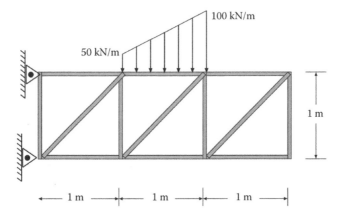

FIGURE 2.18 Beam structure for Problem 2.7.

2.7 For beam-truss structure shown in Figure 2.18, determine the maximum nodal displacement. Give $E = 180\,\text{GPa}$, $A = 3.25 \times 10^{-2}$, and $I_{zz} = 3.5 \times 10^{-4}\,\text{m}^4$.

2.8 For the beam truss shown in Figure 2.19, a transient force, with a frequency $f = 0.02\,\text{Hz}$ and an amplitude Ap $= 50\,\text{kN}$, is oscillating harmonically around a steady load Fs $= 25\,\text{kN}$. Use ANSYS to determine the displacement at the point where the force is applied as a function of time. The total time for the loading process is $250\,\text{s}$. Use the given expression for the force to solve the problem. Given $E = 210\,\text{GPa}$, $A = 2.5 \times 10^{-2}\,\text{m}^2$, and $I_{zz} = 4.5 \times 10^{-4}\,\text{m}^4$.

$$F(t) = \text{Fs} + \text{Ap}\sin(2\pi f t)$$

2.9 For the beam truss shown in Figure 2.20, transient forces at two locations are oscillating harmonically. Use ANSYS to determine the displacement at the points

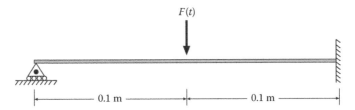

FIGURE 2.19   Beam truss under dynamics loading.

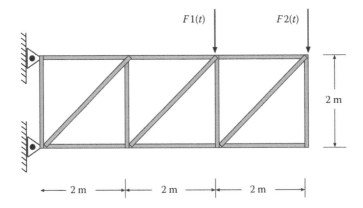

FIGURE 2.20   Beam truss under dynamics loading at two locations.

where the forces are applied as a function of time. The total time for the loading process is 300 s. Use the given sinusoidal expressions to solve the problem. Given $E = 180\,\text{GPa}$, $A = 3.5 \times 10^{-2}\,\text{m}^2$, and $I_{zz} = 5 \times 10^{-4}\,\text{m}^4$.

$$F1(t) = 10 \times 10^3 \sin(0.01\pi t)$$

$$F2(t) = 20 \times 10^3 \sin(0.015\pi t)$$

3 Solid Mechanics and Vibration

3.1 DEVELOPMENT OF PLANE STRESS–STRAIN ELEMENTS

In this chapter, finite element development of two-dimensional solid elements is described. The development is limited to plane stress–strain elements. The plane stress is defined as the state of stress in which the normal and shear stresses are perpendicular to the plane, and is assumed zero. Or, the loads on the body are in the x–y plane only. On the other hand, when the strain is normal to the x–y plane, the strain is called plane strain. Physically, the plane stress–strain elements are characterized by

$$\sigma_z = 0, \quad \tau_{yz} = 0, \quad \tau_{xz} = 0 \tag{3.1}$$

Figure 3.1 shows a two-dimensional stress acting on an element with length dx and width dy. The element is treated as two-dimensional with a unit depth. The normal stresses σ_x and σ_y are acting in the x- and y-directions, respectively. Shear stress τ_{xy} is acting in the y-direction normal to x-plane, and shear stress τ_{yx} is acting in the x-direction normal to y-plane. From the moment equilibrium, τ_{xy} must be equal to τ_{yx}.

The three independent shear stresses can be presented in a vector form as follows:

$$\{\underline{\sigma}\} = \begin{Bmatrix} \sigma_x \\ \sigma_y \\ \tau_{xy} \end{Bmatrix} \tag{3.2}$$

The principal stresses are the maximum and minimum stresses in the two-dimensional plane. The principal stresses can be calculated using the following expressions:

$$\sigma_1 = \sigma_{max} = \frac{\sigma_x + \sigma_y}{2} + \sqrt{\left(\frac{\sigma_x - \sigma_y}{2}\right)^2 + \tau_{xy}^2} \tag{3.3}$$

$$\sigma_2 = \sigma_{min} = \frac{\sigma_x + \sigma_y}{2} - \sqrt{\left(\frac{\sigma_x - \sigma_y}{2}\right)^2 + \tau_{xy}^2} \tag{3.4}$$

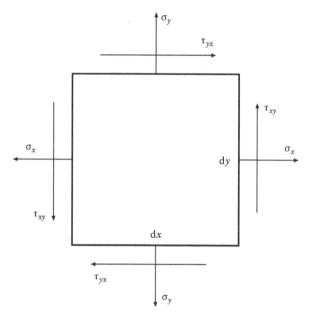

FIGURE 3.1   Two-dimensional state of stress.

The principal angle is defined as the normal whose direction is perpendicular to the plane on which the maximum or minimum principal stress acts. The principal angle can be calculated using the following expression:

$$\theta_p = \frac{1}{2}\tan^{-1}\left(\frac{2\tau_{xy}}{\sigma_x - \sigma_y}\right) \tag{3.5}$$

Figure 3.2 shows the principal stresses, and their direction, acting on a solid element. The strain vector is given by

$$\{\epsilon\} = \begin{Bmatrix} \epsilon_x \\ \epsilon_y \\ \gamma_{xy} \end{Bmatrix} = \begin{Bmatrix} \dfrac{\partial u}{\partial x} \\ \dfrac{\partial v}{\partial x} \\ \dfrac{\partial u}{\partial y} + \dfrac{\partial v}{\partial x} \end{Bmatrix} \tag{3.6}$$

From Hooke's law, the normal and shear strains can be calculated for a two-dimensional solid material using the following expression:

$$\epsilon_x = \frac{\sigma_x}{E} - v\frac{\sigma_y}{E} \tag{3.7}$$

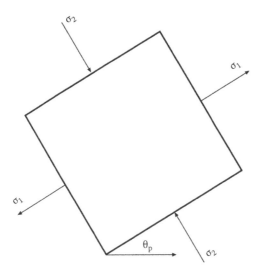

FIGURE 3.2 Principal stresses and the direction.

$$\epsilon_y = \frac{\sigma_y}{E} - v\frac{\sigma_x}{E} \tag{3.8}$$

$$\gamma_{xy} = \frac{E}{2(1-v)}\tau_{xy} \tag{3.9}$$

where v is the Poisson's ratio, which is defined as lateral strain over the axial strain. Normal and shear stresses in Equation 3.2 can be expressed in the following form:

$$\begin{Bmatrix} \sigma_x \\ \sigma_y \\ \tau_{xy} \end{Bmatrix} = \frac{E}{1-v^2}\begin{bmatrix} 1 & v & 0 \\ v & 1 & 0 \\ 0 & 0 & \frac{1-v}{2} \end{bmatrix}\begin{Bmatrix} \epsilon_x \\ \epsilon_y \\ \gamma_{xy} \end{Bmatrix} \tag{3.10}$$

Or, in a compact form,

$$\{\underline{\sigma}\} = \{\underline{D}\}\{\underline{\epsilon}\} \tag{3.11}$$

The first step for the finite element formulation is to divide the domain into elements. Different types of elements can be used, and the simplest type is the linear triangular elements.

Figure 3.3 shows a plate subjected to tensile stress, and the domain is divided into linear triangular elements. The linear triangular elements have three nodes.

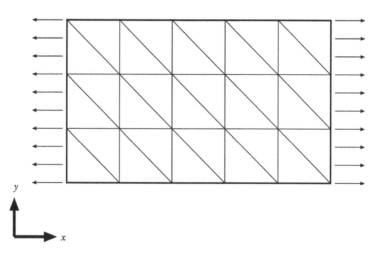

FIGURE 3.3 Finite element mesh for a plate.

The nodes are named i, j, and m. At each node, there are 2 degrees of freedom, displacements in x and y, as shown in Figure 3.4. The nodal displacement vector $\{\underline{d}\}$ is given by

$$\{\underline{d}\} = \begin{Bmatrix} \underline{d_i} \\ \underline{d_j} \\ \underline{d_m} \end{Bmatrix} = \begin{Bmatrix} u_i \\ v_i \\ u_j \\ v_j \\ u_m \\ v_m \end{Bmatrix} \tag{3.12}$$

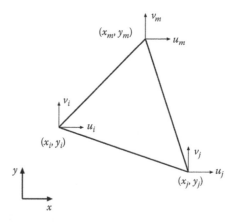

FIGURE 3.4 Triangular element with displacements at the nodes.

Therefore, there are 6 degrees of freedom for each element. Since it is a linear element, linear displacement functions can be selected as

$$u(x, y) = a_1 + a_2 x + a_3 y \tag{3.13}$$

$$v(x, y) = a_4 + a_5 x + a_6 y \tag{3.14}$$

The displacement equations can be solved because there are six constants, a_1 to a_6, and six equations. The nodal x- and y-displacements for all nodes can be expressed in the following matrix form:

$$\begin{Bmatrix} u_i \\ u_j \\ u_m \end{Bmatrix} = \begin{bmatrix} 1 & x_i & y_i \\ 1 & x_j & y_j \\ 1 & x_m & y_m \end{bmatrix} \begin{Bmatrix} a_1 \\ a_2 \\ a_3 \end{Bmatrix} \tag{3.15}$$

$$\begin{Bmatrix} v_i \\ v_j \\ v_m \end{Bmatrix} = \begin{bmatrix} 1 & x_i & y_i \\ 1 & x_j & y_j \\ 1 & x_m & y_m \end{bmatrix} \begin{Bmatrix} a_4 \\ a_5 \\ a_6 \end{Bmatrix} \tag{3.16}$$

Solving for a's:

$$\begin{Bmatrix} a_1 \\ a_2 \\ a_3 \end{Bmatrix} = \frac{1}{2A} \begin{bmatrix} \alpha_i & \alpha_j & \alpha_m \\ \beta_i & \beta_j & \beta_m \\ \gamma_i & \gamma_j & \gamma_m \end{bmatrix} \begin{Bmatrix} u_i \\ u_j \\ u_m \end{Bmatrix} \tag{3.17}$$

and

$$\begin{Bmatrix} a_4 \\ a_5 \\ a_6 \end{Bmatrix} = \frac{1}{2A} \begin{bmatrix} \alpha_i & \alpha_j & \alpha_m \\ \beta_i & \beta_j & \beta_m \\ \gamma_i & \gamma_j & \gamma_m \end{bmatrix} \begin{Bmatrix} v_i \\ v_j \\ v_m \end{Bmatrix} \tag{3.18}$$

where A is the area of the element, which is

$$A = \frac{1}{2}[x_i(y_j - y_m) + x_j(y_m - y_i) + x_m(y_i - y_j)] \tag{3.19}$$

and

$$\alpha_i = x_j y_m - y_j x_m, \quad \alpha_j = x_m y_i - y_m x_i, \quad \alpha_m = x_i y_j - y_i x_j$$
$$\beta_i = y_j - y_m, \quad \beta_j = y_m - y_i, \quad \beta_m = y_i - y_j \tag{3.20}$$
$$\gamma_i = x_m - x_j, \quad \gamma_j = x_i - x_m, \quad \gamma_m = x_j - x_i$$

Substituting the values of α's, β's, and γ's into Equations 3.13 and 3.14:

$$u(x, y) = \frac{1}{2A}[(\alpha_i + \beta_i x + \gamma_i y)u_i + (\alpha_j + \beta_j x + \gamma_j y)u_j + (\alpha_m + \beta_m x + \gamma_m y)u_m] \tag{3.21}$$

$$v(x, y) = \frac{1}{2A}[(\alpha_i + \beta_i x + \gamma_i y)v_i + (\alpha_j + \beta_j x + \gamma_j y)v_j + (\alpha_m + \beta_m x + \gamma_m y)v_m] \tag{3.22}$$

The strain vector for a two-dimensional element is given in Equation 3.6. Using the displacements v and u, the strain vector can be expressed as

$$\{\underline{\epsilon}\} = \begin{Bmatrix} \tau_x \\ \tau_y \\ \gamma_{xy} \end{Bmatrix} = \frac{1}{2A} \begin{Bmatrix} \beta_i u_i + \beta_j u_j + \beta_m u_m \\ \gamma_i v_i + \gamma_j v_j + \gamma_m v_m \\ \gamma_i u_i + \beta_i v_i + \gamma_j u_j + \beta_j v_j + \gamma_m u_m + \beta_m v_m \end{Bmatrix} \tag{3.23}$$

The strain vector can be expressed in a matrix form as follows:

$$\{\underline{\epsilon}\} = \frac{1}{2A} \begin{bmatrix} \beta_i & 0 & \beta_j & 0 & \beta_m & 0 \\ 0 & \gamma_i & 0 & \gamma_j & 0 & \gamma_m \\ \gamma_i & \beta_i & \gamma_j & \beta_j & \gamma_m & \beta_m \end{bmatrix} \begin{Bmatrix} u_i \\ v_i \\ u_j \\ v_j \\ u_m \\ v_m \end{Bmatrix} \tag{3.24}$$

The vector (Equation 3.24) can be expressed in a more compact form where the $[\underline{B}]$ matrix is defined as follows:

$$\{\underline{\epsilon}\} = \begin{bmatrix} B_i & B_j & B_m \end{bmatrix} \begin{Bmatrix} \underline{d}_i \\ \underline{d}_j \\ \underline{d}_m \end{Bmatrix} \tag{3.25}$$

where

$$[B_i] = \frac{1}{2A}\begin{bmatrix} \beta_i & 0 \\ 0 & \gamma_i \\ \gamma_i & \beta_i \end{bmatrix}, \quad [B_j] = \frac{1}{2A}\begin{bmatrix} \beta_j & 0 \\ 0 & \gamma_j \\ \gamma_j & \beta_j \end{bmatrix}, \quad [B_m] = \frac{1}{2A}\begin{bmatrix} \beta_m & 0 \\ 0 & \gamma_m \\ \gamma_m & \beta_m \end{bmatrix} \quad (3.26)$$

The matrix (Equation 3.25) can be expressed as

$$\{\underline{\epsilon}\} = [B]\{d\} \tag{3.27}$$

The stress–strain relationship shown in Equation 3.11 can also be expressed using [B] and [D] as follows:

$$\{\underline{\sigma}\} = [D][B]\{d\} \tag{3.28}$$

Using the principle of the minimum potential energy theory, the [k] matrix can be obtained

$$[k] = tA[B]^T[D][B] \tag{3.29}$$

where
   t is the thickness of the element
   A is the area of the element

The global stiffness matrix can be obtained by assembling the stiffness matrix for each element as follows

$$[K] = \sum_{e=1}^{N}[k^{(e)}] \tag{3.30}$$

and the nodal forces are also assembled to form a global force vector as follows

$$[F] = \sum_{e=1}^{N}[f^{(e)}] \tag{3.31}$$

Also, the global displacement vector can be obtained by

$$[d] = \sum_{e=1}^{N}[d^{(e)}] \tag{3.32}$$

Finally,

$$\{F\} = [K]\{d\} \tag{3.33}$$

In order to determine the nodal displacements, the global stiffness matrix must be formulated. The general steps for solving solid mechanics problems are as follows:

a. Divide the domain into elements
b. Formulate the $[B]$ matrix for each element
c. Formulate the $[D]$ matrix for each element
d. Formulate the $[k]$ matrix for each element using calculated $[B]$ and $[D]$ in steps b and c
e. Assemble the $[k]$ matrices to create the $[k]$ matrix
f. Formulate the equation matrix $\{F\} = [K]\{d\}$
g. Apply the boundary conditions to the equation matrix in step f
h. Solve equation matrix in step g to determine the displacements

Example 3.1: A rectangular plate subjected to a force

Determine the nodal displacements for a thin plate subjected to a force equal to 40 kN, as shown in Figure 3.5. Use $E = 210$ GPa, $v = 0.3$, and $t = 0.005$ m. Divide the plate into two elements only, as illustrated in Figure 3.6.

First, the $[B]$, $[D]$, and $[k]$ matrices for element 1 are formulated. The element 1 has the coordinate shown in Figure 3.7, and the nodes are named i, j, and m.

The values of β and γ are required for the $[B]$ matrix using Equation 3.20, and A is calculated from Equation 3.19, we have

$$\beta_i = y_j - y_m = 0 - 0.25 = -0.25$$

$$\beta_j = y_m - y_i = 0.25 - 0 = 0.25$$

$$\beta_m = y_i - y_j = 0 - 0 = 0$$

FIGURE 3.5   Geometry for Example 3.1.

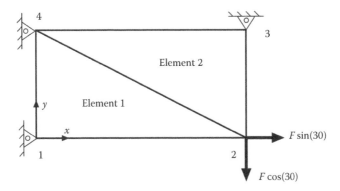

FIGURE 3.6 Element distribution for Example 3.1.

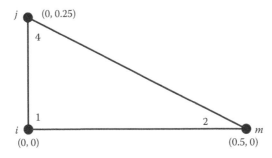

FIGURE 3.7 Element number 1.

$$\gamma_i = x_m - x_j = 0 - 0.5 = -0.5$$

$$\gamma_j = x_i - x_m = 0 - 0 = 0$$

$$\gamma_m = x_j - x_i = 0.5 - 0 = 0.5$$

$$2A = 0.125 \text{ m}^2$$

Then, the [B] matrix is formulated using Equation 3.24

$$[B] = \frac{1}{0.125} \begin{bmatrix} -0.25 & 0 & 0.25 & 0 & 0 & 0 \\ 0 & -0.5 & 0 & 0 & 0 & 0.5 \\ -0.5 & -0.25 & 0 & 0.25 & 0.5 & 0 \end{bmatrix}$$

and the [D] matrix is formulated using Equation 3.11

$$[D] = \frac{210 \times 10^9}{1-3^2} \begin{bmatrix} 1 & 0.3 & 0 \\ 0.3 & 1 & 0 \\ 0 & 0 & \dfrac{1-0.3}{2} \end{bmatrix}$$

Finally, the [k] matrix for element 1 is obtained using Equation 3.29:

$$[k^{(1)}] = 10^9 \begin{bmatrix} 0.6923 & 0.375 & -0.288 & -0.202 & -0.404 & -0.173 \\ 0.375 & 1.255 & -0.173 & -0.101 & -0.202 & -1.15 \\ -0.288 & -0.173 & 0.288 & 0 & 0 & 0.173 \\ -0.202 & -0.101 & 0 & 0.101 & 0.202 & 0 \\ -0.404 & -0.202 & 0 & 0.202 & 0.404 & 0 \\ -0.173 & -1.15 & 0.173 & 0 & 0 & 1.153 \end{bmatrix}$$

A similar procedure is done for element number 2, as shown in Figure 3.8. First, the [B], [D], and [k] matrices for element 2 are formulated. The nodes are named i, j, and m, as shown.

The values of β's and γ's are required for the [B] matrix, using Equation 3.20, and A is calculated from Equation 3.19, we have

$$\beta_i = y_j - y_m = 0.25 - 0.25 = 0$$

$$\beta_j = y_m - y_i = 0.25 - 0 = 0.25$$

$$\beta_m = y_i - y_j = 0 - 0.25 = -0.25$$

$$\gamma_i = x_m - x_j = 0 - 0.5 = -0.5$$

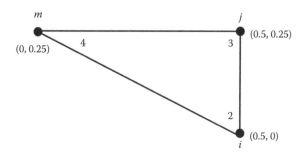

FIGURE 3.8   Element number 2.

$$\gamma_j = x_i - x_m = 0.5 - 0 = 0.5$$

$$\gamma_m = x_j - x_i = 0.5 - 0.5 = 0$$

$$2A = 0.125 \text{ m}^2$$

Then, the [B] matrix is formulated using Equation 3.24

$$[B] = \frac{1}{0.125} \begin{bmatrix} 0 & 0 & 0.25 & 0 & -0.25 & 0 \\ 0 & -0.5 & 0 & 0.5 & 0 & 0 \\ -0.5 & 0 & 0.25 & 0.25 & 0 & -0.25 \end{bmatrix}$$

and the [D] matrix is formulated using Equation 3.11

$$[D] = \frac{210 \times 10^9}{1 - 3^2} \begin{bmatrix} 1 & 0.3 & 0 \\ 0.3 & 1 & 0 \\ 0 & 0 & \dfrac{1-0.3}{2} \end{bmatrix}$$

Finally, the [k] matrix for element 2 is obtained using Equation 3.29:

$$[k^{(2)}] = 10^9 \begin{bmatrix} 0.404 & 0 & -0.404 & -0.202 & 0 & 0.202 \\ 0 & 1.15 & -0.173 & -0.15 & 0.173 & 0 \\ -0.404 & -0.173 & 0.692 & 0.373 & -0.288 & -0.202 \\ -0.202 & -0.15 & 0.375 & 1.254 & -0.173 & -0.101 \\ 0 & 0.173 & -0.288 & -0.1731 & 0.288 & 0 \\ 0.202 & 0 & -0.202 & -0.101 & 0 & 0.101 \end{bmatrix}$$

Using Equation 3.30 the global matrix is formulated as

$$[K] = 10^9 \begin{bmatrix} 0.6923 & 0.3750 & -0.288 & -0.202 & 0 & 0 & -0.404 & -0.173 \\ 0.3750 & 1.255 & -0.173 & -0.101 & 0 & 0 & -0.202 & -1.15 \\ -0.288 & -0.173 & 0.692 & 0 & -0.404 & -0.202 & 0 & 0.375 \\ -0.202 & -0.101 & 0 & 1.251 & -0.173 & -1.15 & 0.373 & 0 \\ 0 & 0 & -0.404 & -0.173 & 0.692 & 0.375 & -0.288 & -0.202 \\ 0 & 0 & -0.202 & -1.15 & 0.375 & 1.254 & -0.173 & -0.101 \\ -0.404 & -0.202 & 0 & 0.375 & -0.288 & -0.173 & 0.692 & 0 \\ -0.173 & -1.150 & 0.375 & 0 & -0.202 & -0.101 & 0 & 1.254 \end{bmatrix}$$

The force–displacement equation is assembled using $\{F\} = [K]\{d\}$ as follows:

$$\begin{Bmatrix} f_{1x} \\ f_{1y} \\ f_{2x} = 40 \times 10^3 \sin(30) \\ f_{2y} = -40 \times 10^3 \cos(30) \\ f_{3x} \\ f_{3y} \\ f_{4x} \\ f_{4y} \end{Bmatrix} = [K] \begin{Bmatrix} d_{1x} = 0 \\ d_{1y} = 0 \\ d_{2x} \\ d_{2y} \\ d_{3x} = 0 \\ d_{3y} = 0 \\ d_{4x} = 0 \\ d_{4y} = 0 \end{Bmatrix}$$

Simplifying the above relationship, we have

$$10^9 \begin{bmatrix} 0.692 & 0 \\ 0 & 1.251 \end{bmatrix} \begin{Bmatrix} d_{2x} \\ d_{2y} \end{Bmatrix} = 10^3 \begin{Bmatrix} 40 \sin 30 \\ -40 \cos 30 \end{Bmatrix}$$

Finally, solving the above equation, we have

$$d_{2x} = 0.289 \times 10^{-4} \text{ m}$$

$$d_{2y} = -0.276 \times 10^{-4} \text{ m}$$

As expected, the negative d_{2y} indicates that the node 2 is moved downward, and positive d_{2x} indicates that the node 2 is moved to the right.

3.2   STRESS CONCENTRATION OF A PLATE WITH HOLE

For the geometry in Figure 3.9, use the ANSYS to determine the maximum stress in the x-direction. Also, compare the ANSYS result with maximum stress using the stress concentration factor, as shown in Figure 3.11. The applied pressure is 1 kN/m^2, and let $E = 210 \text{ GPa}$, $v = 0.25$, and $t = 0.005 \text{ m}$.

In Figure 3.10a, a plate with a hole is subjected to a tensile pressure at both the left and right sides. The stress near the hole is typically assumed uniform. However, from the experimental observations, the stress is not uniform and it has a maximum value near the hole, and it is greater than the average stress, as shown in Figure 3.10b.

The complexity of maximum stress can be conveniently treated using the stress concentration factor (K). The maximum stress is equal to the average stress multiplied by the stress concentration factor (K). The maximum stress can be obtained from the following relationship:

$$\sigma_{max} = K\sigma_{ave} \tag{3.34}$$

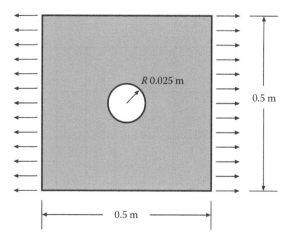

FIGURE 3.9 A plate with a hole subjected to a tensile pressure.

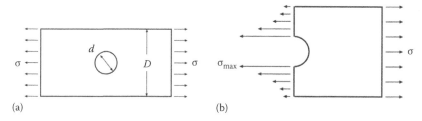

(a) (b)

FIGURE 3.10 (a) A plate with hole under stress, and (b) stress distribution near the hole.

The stress concentration for a solid plate with a hole can be obtained using Figure 3.11. The x-axis represents the ratio of the diameter of the hole to the width of the plate, while the y-axis is the corresponding stress concentration factor. Note that decreasing the diameter of the hole increases the stress concentration up to three times the average stress.

Another practical geometry is a block subjected to concentrated force. In order to determine the stress in the block, the stress at any vertical location can be calculated using $\sigma = F/A$. However, the experimental observations indicate that the result will not be correct. The strain is maxima near the applied force. Therefore, the corresponding stress must be maxima. The stress is maxima close to the section of the applied force, and becomes nearly uniform at the center of the block.

This example is limited to structural analysis. Hence, select Structural only in the Preferences. A solid element is used, and its shape is a triangle with six nodes. The following steps are for selecting the element type.

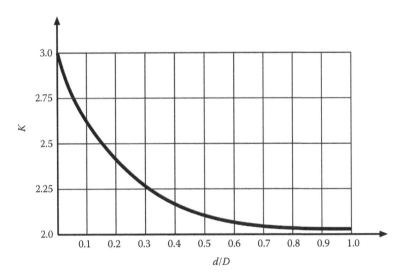

FIGURE 3.11   Stress concentration factors for a plate with a hole.

Double click on the ANSYS icon

Main Menu > Preferences

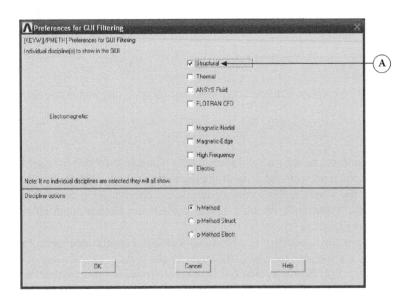

A select the Structural

OK

Main Menu > Preprocessor > Element type > Add/Edit/Delete

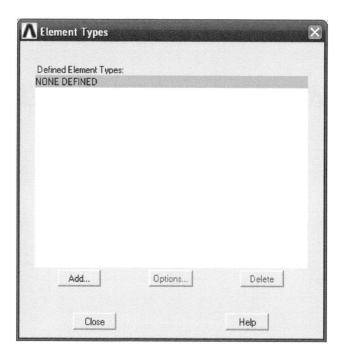

Add...

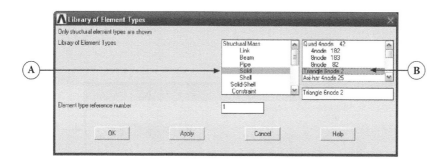

A select Solid

B select Triangle 6node 2

OK

The Triangle 6node 2 is a triangular-shaped element with six nodes. A linear triangular element is not available in the ANSYS element list. A quadratic element with four and eight nodes is also available. The plate has a thickness, in the option, select plane stress with thickness. In the Real constant window, the thickness is specified.

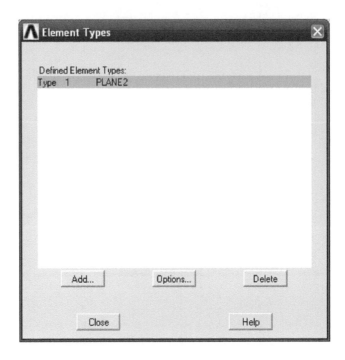

Option

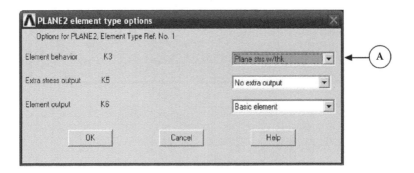

A select Plane strs w/thk

OK

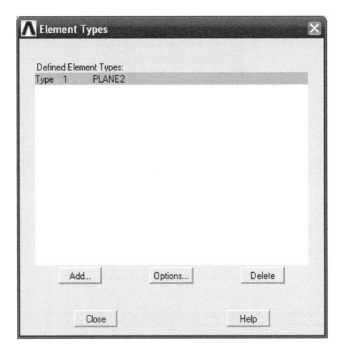

Close

Main Menu > Preprocessor > Real Constants > Add/Edit/Delete

Add

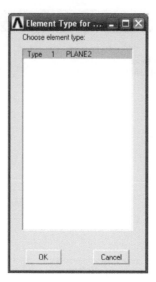

OK

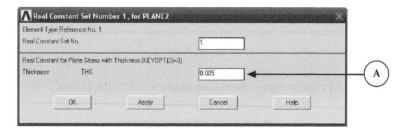

A type 0.005 in Thickness

OK

Close

For the material properties, the modulus of elasticity and Poisson's ratio are required to solve the problem.

Main Menu > Preprocessor > Material Props > Material Models

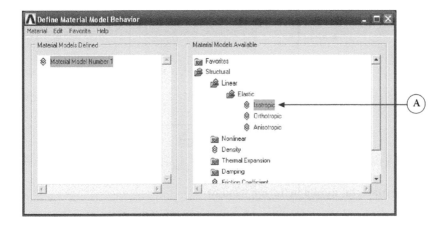

A Double click on Structure > Linear > Elastic > Isotropic

The following windows will show up

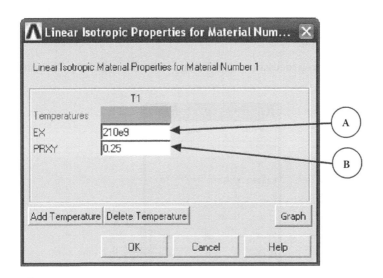

A type 210e9 in EX

B type 0.25 in the PRXY

OK

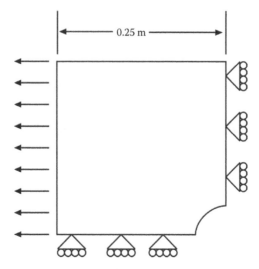

FIGURE 3.12   Symmetry boundary conditions for a plate with hole.

Close the material model behavior window

The geometry is modeled by creating a square and a circle. A Boolean operation is utilized to remove the circle from the square using overlap and delete commands. Alternatively, the circle can directly be removed using the subtract command. The advantage of symmetry in the problem is considered. Only the upper right quarter is considered. Figure 3.12 shows the considered geometry used to solve the problem. Notice that the imposed boundary condition at the vertical symmetry line is zero displacement in the x-direction, and zero displacement in the y-direction at the horizontal symmetry line.

Main Menu > Preprocessor > Modeling > Create > Area > Rectangle > By 2 Corners

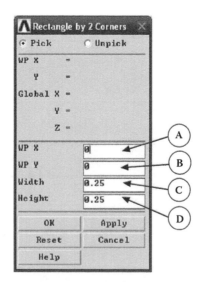

A type 0 in WP X

B type 0 in WP Y

C type 0.25 in Width

D type 0.25 in Height

OK

Main Menu > Preprocessor > Modeling > Create > Area > Circle > Solid Circle

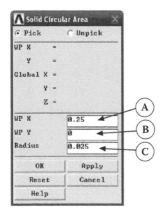

A type 0.25 in WP X

B type 0.0 in WP Y

C type 0.025 in Radius

OK

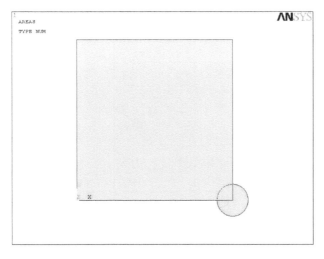

ANSYS graphics shows the square and the circle

Main Menu > Preprocessor > Modeling > Operate > Booleans > Overlap > Areas

Click on to select all areas

Main Menu > Preprocessor > Modeling > Delete > Area and below

Click on the hole to highlight it

OK

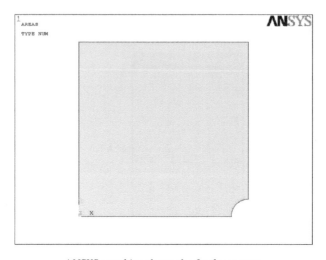

ANSYS graphics shows the final geometry

The geometry is meshed with the triangular six-node elements. A free mesh is generated using the smart mesh option. The mesh refinement is 1.

Main Menu > Preprocessor > Meshing > Mesh Tool

A select Smart Size

B set Smart Size to 1

C Mesh

Click on **Pick All** to mesh the computational domain

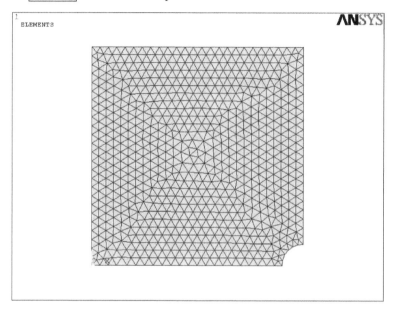

ANSYS graphics shows the mesh

Pressure is applied at the left vertical side of the geometry. A negative pressure means that the pressure is tensile.

Main Menu > Solution > Define Load > Apply > Structural > Pressure > On Lines

In the ANSYS graphics, click on the left vertical line.

OK

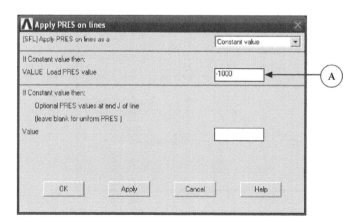

A type -1000 in Load PRES value

OK

Main Menu > Solution > Define Load > Apply > Structural > Displacement > On Lines

In the ANSYS graphics, click on the right vertical line, where zero *x*-direction displacement is applied.

OK

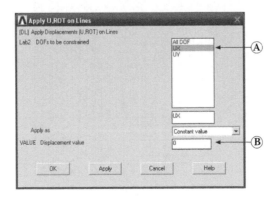

A select UX

B type 0 in Displacement value

OK

Main Menu > Solution > Define Load > Apply > Structural > Displacement > On Lines

In the ANSYS graphics, click on the horizontal bottom line, where zero *y*-direction displacement is applied.

OK

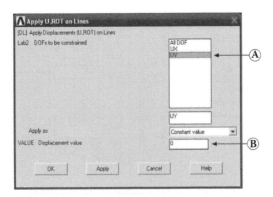

A select UY

B type 0 in Displacement value

The final step is to run the solution. ANSYS will assemble the stiffness matrix, apply the boundary conditions, and solve the stiffness matrix. Results can be plotted and listed in the general preprocessor task.

Main Menu > Solution > Solve > Current LS

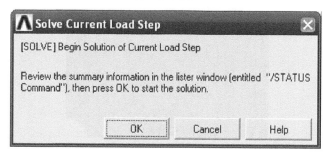

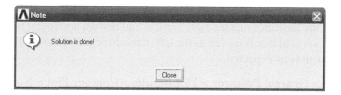

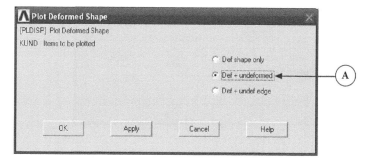

The above windows indicate that the solution task is completed successfully. The next step is for getting the results. The first figure shows the plate before and after deformation. Inspecting this figure is very important to determine if the problem is solved correctly. The second figure is the stress contours in the x-direction. This figure is used to determine the maximum stress in the x-direction.

Main Menu > General Postproc > Plot Results > Deformed Shape

A select Def + undeformed

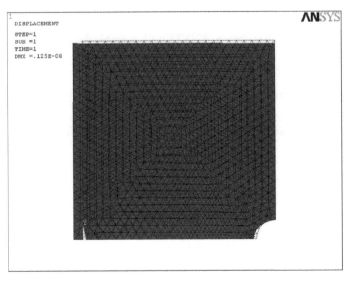

ANSYS graphics shows the plate before and after deformation

The above figure indicates that the right vertical and bottom horizontal line are fixed, while the left vertical line is moved to the left in the direction of the applied pressure. The initial result is as expected.

Main Menu > General Postproc > Plot Results > Contour Plot > Nodal Solu

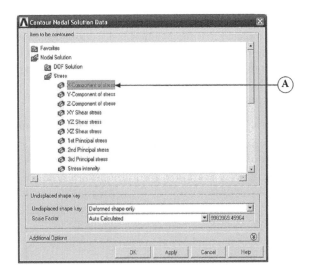

A click on Nodal Solution > Stress > X-Component of stress

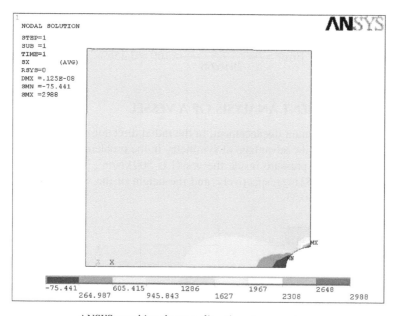

ANSYS graphics shows x-direction stress contours

The stress contours show the location of the maximum stress in the *x*-direction, as expected, at the top of the hole. The accuracy of the result can additionally be improved if the number of element in the model is increased. The size of the geometry could be increased by four if the symmetry is not considered, which will significantly increase the numerical errors.

As shown in the stress contours, the maximum stress is 2988 N/m². The maximum stress can also be obtained using the stress concentration in Figure 3.11. The average stress is

$$\sigma_{ave} = \frac{1000 \times 0.5}{0.5 - 0.05} = 1111.11 \, \text{N/m}^2$$

Using Figure 3.11, the stress concentration should be

$$K = 2.625$$

and the maximum stress can be calculated using Equation 3.34 as follows:

$$\sigma_{max} = K\sigma_{ave} = 2.625 \times 1111.11 = 2916.66 \text{ N/m}^2$$

Comparing the ANSYS maximum stress with maximum stress obtained from the figure, the difference between the two results is extremely small, and the error is equal to

$$\text{Error} = \frac{2988 - 2916.66}{2916.66} \times 100 = 2.456\%$$

3.3   DISPLACEMENT ANALYSIS OF A VESSEL

Determine the maximum displacement, in the radial direction, of the vessel shown in Figure 3.13. Take the advantage of symmetry in the problem to reduce the computational size. The pressure inside the vessel is 500 kN/m^2. The inner and outer radii are 0.5 and 0.52 m, respectively, and the height of the vessel is 0.25 m. Let $E = 180 \text{ GPa}$ and $v = 0.33$.

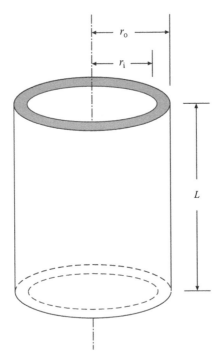

FIGURE 3.13   A vessel subjected to a pressure at the inner surface.

Double click on the ANSYS icon

Main Menu > Preferences

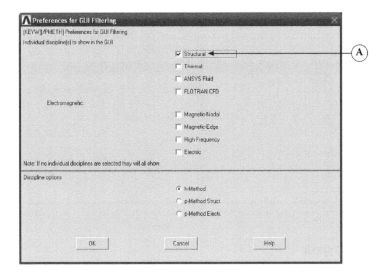

A select the Structural

OK

This example is limited to structural analysis. Hence, select Structural only. A solid element is used, and its shape is a rectangle with four nodes. The following steps are for selecting the element type and its behavior.

Main Menu > Preprocessor > Element type > Add/Edit/Delete

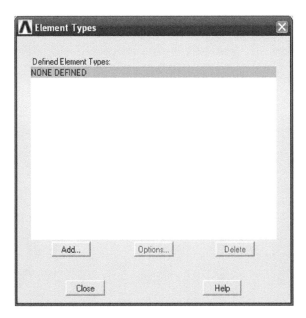

Add...

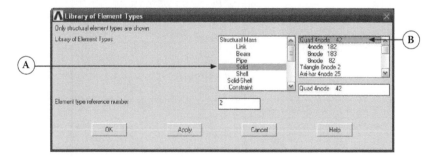

A select Solid

B select Quad 4node 42

OK

In the element type, Quad means that a quadratic element is selected. The first digit of number 42 is the number of nodes in the element, and the second digit is the number of degrees of freedom in each node, which is the x- and y-displacements. In the following steps, the element behavior is converted from plane stress to axisymmetric. The results from the axisymmetric analysis should be more accurate than those from the three-dimensional analysis, and the computational time should be less.

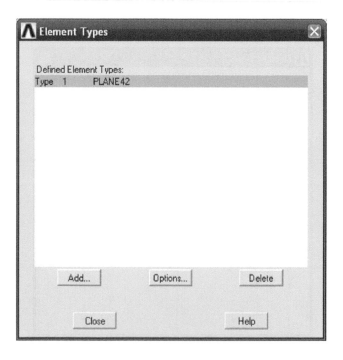

Option

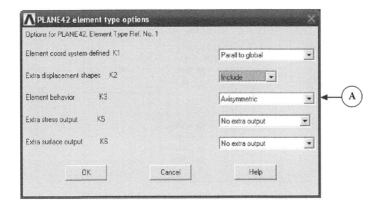

A select Axisymmetric in the Element behavior

OK

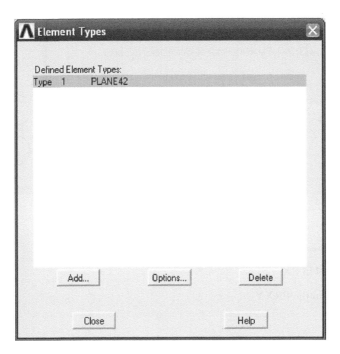

Close

For the material properties, the elasticity and Poisson's ratio are required to solve the problem.

Finite Element Simulations Using ANSYS

Main Menu > Preprocessor > Material Props > Material Models

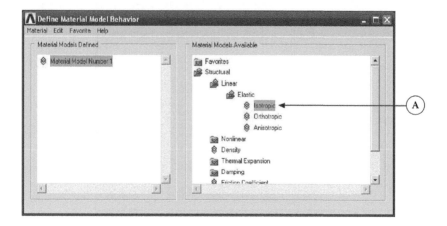

A double click on Structure > Linear > Elastic > Isotropic

The following widows will show up

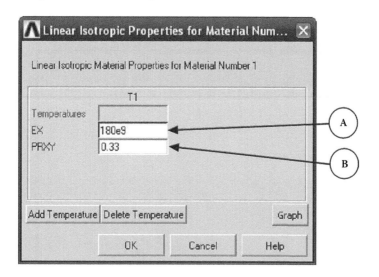

A type 180e9 in EX

B type 0.33 in the PRXY

Close the material model behavior window

The geometry is simply a rectangle. The advantage of symmetry in the problem is considered. Only a cross-section area of the vessel is considered. Figure 3.14 shows the considered geometry used to solve the problem. The axis of rotation must be the y-axis, and no geometry is allowed in the negative x-axis.

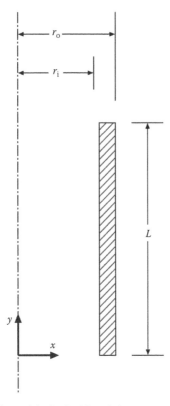

FIGURE 3.14   Axisymmetric models for Problem 3.4.

Main Menu > Preprocessor > Modeling > Create > Areas > Rectangle > By 2 Corners

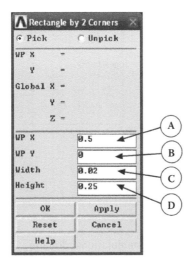

A type 0.5 in WP X

B type 0 in WP Y

C type 0.02 in Width

D type 0.25 in the Height

Main Menu > Preprocessor > Meshing > Mesh Tool

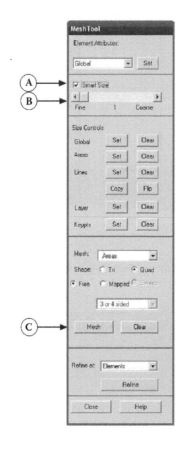

A select Smart Size

B set Smart Size to 1

C Mesh

Click on $\boxed{\textbf{Pick All}}$ to select the computational domain

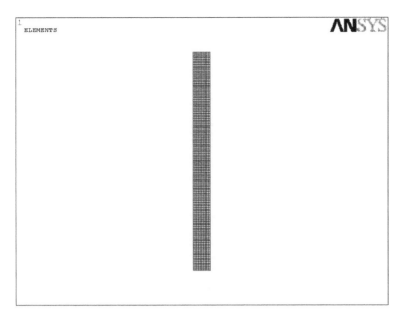

ANSYS graphics shows mesh of the cross-section area of the vessel

Pressure is applied at the left vertical line. A positive pressure means that the pressure is compression, while the top and bottom lines are fixed.

Main Menu > Solution > Define Load > Apply > Structural > Pressure > On Lines

In the ANSYS graphics, click on the left vertical line where pressure is applied.

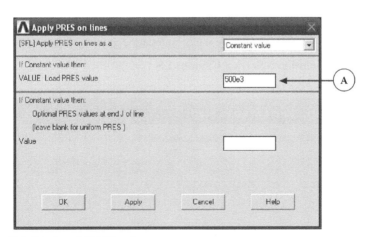

A type 500e3 in Load PRES value

Main Menu > Solution > Define Load > Apply > Structural > Displacement > On Lines

In the ANSYS graphics, click on the top and bottom lines that are fixed.

OK

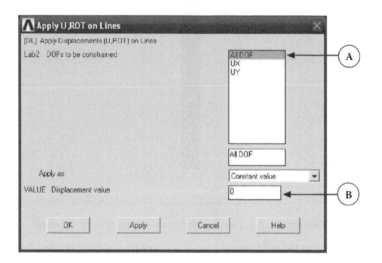

A select All DOF

B type 0 in Displacement value

OK

The final step in the solution task is to run the solution. ANSYS will assemble the stiffness matrix, apply the boundary conditions, and solve the stiffness matrix. Results can be plotted and listed in the General Preprocessor task. Inspecting the deformation plot will help to identify if any boundary condition was wrongly applied.

Main Menu > Solution > Solve > Current LS

OK

Close

The above window indicates that the solution task is accomplished successfully. The next step is for getting the results.

Main Menu > General Postproc > Plot Results > Deformed Shape

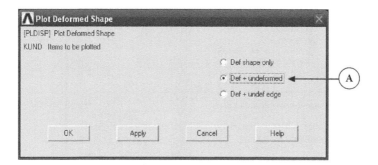

A select Def + undeformed

OK

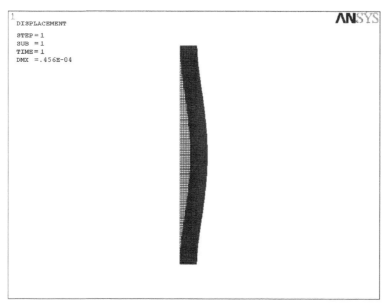

ANSYS graphics shows the vessel wall before and after deformation

The above figure indicates that the upper and lower sides are fixed. At the middle, the vessel wall is bended to the outside, and the maximum displacement is exactly in the middle, as expected.

Main Menu > General Postproc > Plot Results > Contour Plot > Nodal Solu

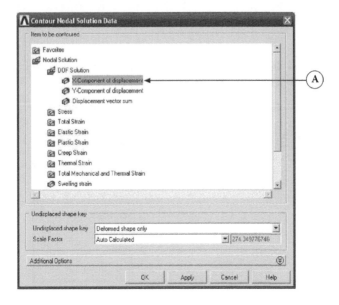

A click on Nodal Solution > DOF > X-Component of displacement

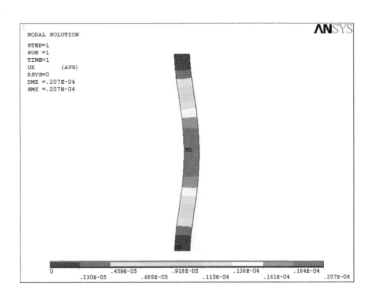

ANSYS graphics shows x-direction displacement contours

The displacement contours shows the location of the maximum displacement in the *r*-direction. Maximum displacement is exactly in the middle. The maximum displacement is equal to 0.456×10^{-4} m. To visualize the results in three-dimensional space, follow the following steps:

Utility Menu > PlotCtrls > Style > Symmetry Expansion > 2D Axi-Symetric

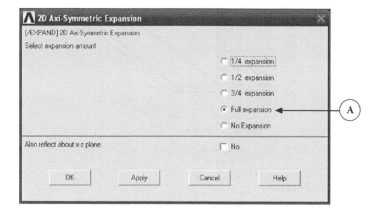

A Full expansion

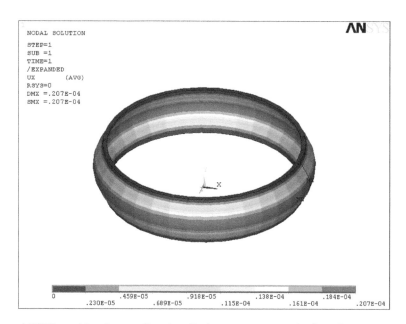

ANSYS graphics shows x-direction displacement contours in three dimensions

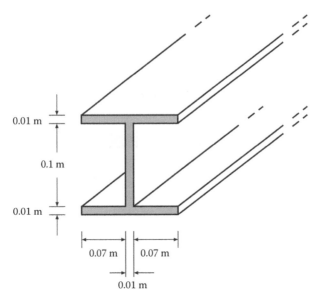

0.01 m

0.1 m

0.01 m

0.07 m   0.07 m

0.01 m

FIGURE 3.15   The I-beam.

3.4   THREE-DIMENSIONAL STRESS ANALYSES OF I-BEAMS

The I-beam shown in Figure 3.15 is fixed at the left side and under a uniform pressure of 500 kPa at its upper surface. The total length of the I-beam is 2 m. Solve the problem in a three-dimensional space to determine the maximum displacement in the y-direction. The properties of the I-beam are $E = 210$ GPa and $v = 0.33$.

This example is limited to structural analysis. Hence, select Structural only. Three-dimensional solid elements must be used, and shape of the selected element is tetrahedral with 10 nodes. The following steps are for selecting the element type.

Double click on the ANSYS icon

Main Menu > Preferences

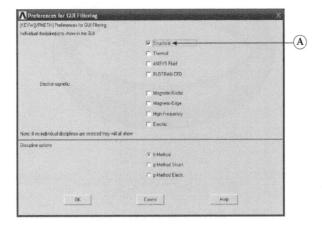

A select the Structural

Main Menu > Preprocessor > Element type > Add/Edit/Delete

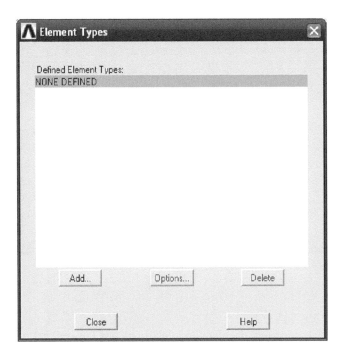

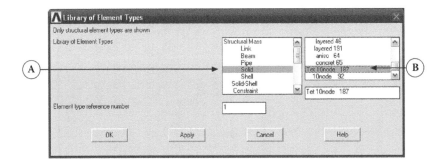

A select Solid

B select Tet 10node 187

OK

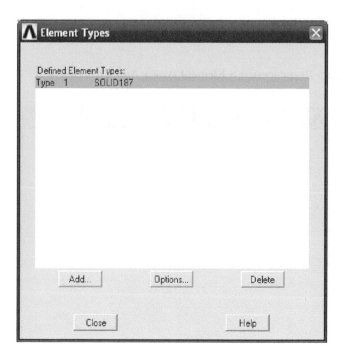

Close

For the material properties, the modulus of elasticity and Poisson's ratio are required to solve the problem.

Main Menu > Preprocessor > Material Props > Material Models

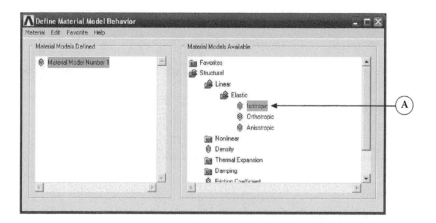

A double click on Structure > Linear > Elastic > Isotropic

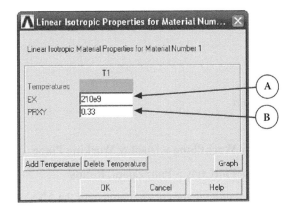

A type 210e9 in EX

B type 0.33 in the PRXY

Close the material model behavior window

The used technique to model the I-beam is quite simple. The following steps are performed to create the I-beam:

1. The work plane is prepared with an appropriate grid spacing and size.
2. Key points are created on the work plane.
3. Lines are created by connecting the two key points to form the frame of the I-beam.
4. The area of cross section of the I-beam is created.
5. The cross-section area of the I-beam is extruded to form a three-dimensional I-beam.

Utility Menu > WorkPlane > WP Setting

A select Grid Only

B type 0.01 in Snap Incr

C type 0.01 in Spacing

D type 0 in Minimum

E type 0.15 in Maximum

ANSYS Utility Menu > WorkPlane > Display Working Plane

ANSYS Utility Menu > PlotCtrls > Pan, Zoom, Rotate

Click on zoom in and out, moving the cursor button until the ANSYS graphics shows all grids.

ANSYS Main Menu > Preprocessor > Modeling > Create > Keypoints > On Working Plane

Using the mouse, click on the ANSYS graphics window at the location key points, as shown in the figure below.

ANSYS graphics shows the key points

ANSYS Main Menu > Preprocessor > Modeling > Create > Lines > Lines > Straight Line

Create a line by clicking on two key points, as shown below

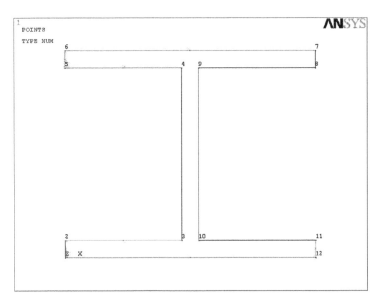

ANSYS graphics shows lines

ANSYS Main Menu > Preprocessor > Modeling > Create > Areas > Arbitrary > By Lines

Click on all lines to create one area

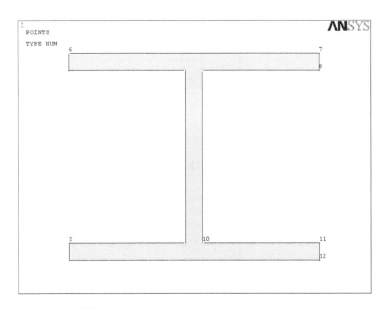

ANSYS graphics shows cross-section area of the I-beam

The cross-section area of the beam is ready to be extruded. The area is extruded along its normal direction in the negative z-axis, and the total length of the beam is 2 m.

ANSYS Main Menu > Preprocessor > Modeling > Operate > Extrude > Areas > Along Normal

Click on the area of the I-beam

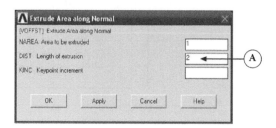

A type 2 in the Length of extrusion

The area is now extruded in the negative z-direction. The isometric view and projection views of the beam can be created using the Pan-Zoom Rotate tool. In addition, the beam can be rotated. The beam is meshed with smart size of 7 to have a fast solution process for this example. The smaller smart mesh size will take a considerable amount of time, but the results will definitely be more accurate.

Utility Menu > PlotCtrls > Pan Zoom Rotate

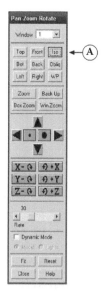

A click on the Iso

Close

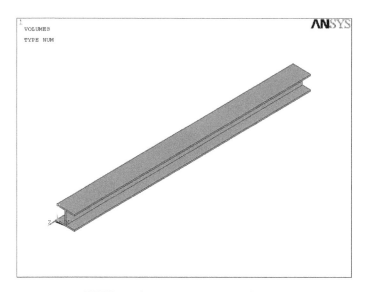

ANSYS graphics isometric view of the beam

Main Menu > Preprocessor > Meshing > Mesh Tool

A select Smart Size

B set Smart Size to 7

C Mesh

Click on **Pick All** in meshed areas window

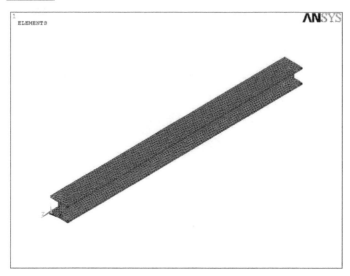

ANSYS graphics shows the mesh

The left area of the beam is fixed, and pressure is applied at the upper surface of the beam. A positive pressure means that the pressure is compression.

Main Menu > Solution > Define Load > Apply > Structural > Displacement > On Areas

In the ANSYS graphics, click on the left area, where zero displacement is applied.

OK

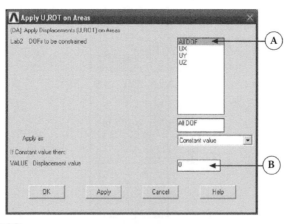

A select All DOF

B type 0 in Displacement value

OK

Main Menu > Solution > Define Load > Apply > Structural > Pressure > On Areas

In the ANSYS graphics, click on the upper surface

OK

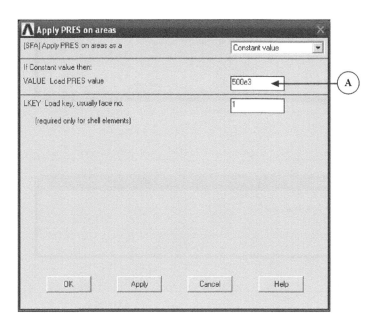

A type 500e3 in Load PRES value

OK

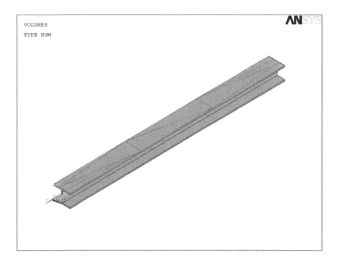

ANSYS graphics shows the applied boundary condition

Main Menu > Solution > Solve > Current LS

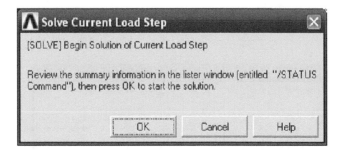

 OK

 Close

The above windows indicate that the solution task is accomplished successfully. The next step is for getting the results.

Main Menu > General Postproc > Plot Results > Deformed Shape

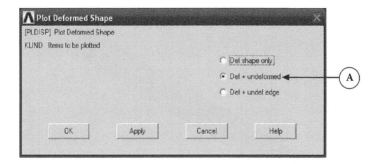

A select Def + undeformed

OK

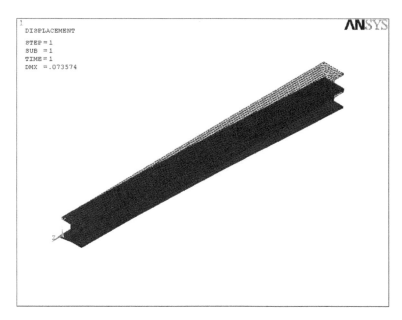

ANSYS graphics shows the beam before and after deformation

Main Menu > General Postproc > Plot Results > Contour Plot > Nodal Solu

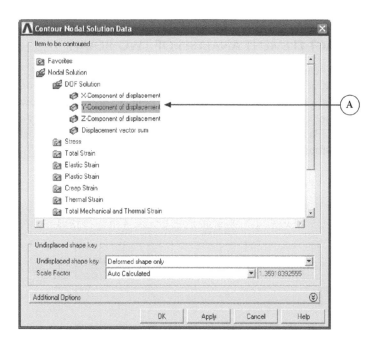

A click on Nodal Solution > DOF Solution > Y-Component of displacement

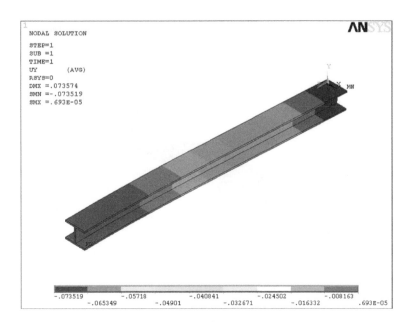

NODAL SOLUTION
STEP=1
SUB =1
TIME=1
UY   (AVG)
RSYS=0
DMX =.073574
SMN =-.073519
SMX =.693E-05

ANSYS graphics shows Y-direction of displacement

As shown, the maximum displacement is 0.0693 m in the negative y-direction. The stresses can be presented in a similar manner.

3.5   CONTACT ELEMENT ANALYSIS OF TWO BEAMS

Two horizontal beams are placed close to each other, as shown in Figure 3.16. Both beams are fixed at the left side, and a pressure of 15 MPa is applied only to the upper beam. As a result, the upper beam will deflect and meet the lower beam. Determine the maximum deflection of the first and second beams in the y-direction. Given $E = 200$ GPa and $v = 0.33$ for both beams. The friction coefficient between the two beams is estimated to be 0.2.

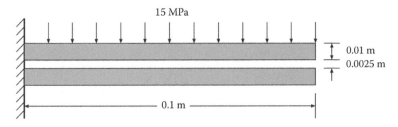

FIGURE 3.16   A contact element problem.

Contact problems are highly nonlinear and require large computer resources to solve. Additionally, the physics behind the contact problem is relatively complex, and it is important to understand the physics of the problem to solve it as accurately as possible. Too many parameters are required, and just using the ANSYS default settings will give a reasonable result. Contact problems present two main difficulties. First, the regions of contact cannot be predicted accurately. The contact region depends on the loads, material, boundary conditions, and other factors. Second, most contact problems need to account for friction. There are several friction laws and models to choose from, and all are nonlinear, which makes solution convergence even more difficult.

The basic steps for performing a contact analysis are as follows:

1. *Create the model geometry and mesh.* This step is typical for solving any ANSYS problem. The solid model is created that represent the geometry, element types, real constants, and material properties.
2. *Identify the contact pairs.* The contact pairs are the region where contact might occur during the deformation of the model. Once contact surfaces are identified, they must be defined via target and contact elements.
3. *Select contact and target surfaces.* Contact elements are constrained against penetrating the target surface, but target elements can penetrate through the contact surface. If a convex surface is expected to meet a flat surface, the convex surface should be the contact surface. If one surface is smaller than the other surface, the smaller surface should be the contact surface.
4. *Define the target surface.* The target line should be defined in this step, and this step can be accomplished effectively using the contact wizard.
5. *Define the contact surface.* The contact line should be defined in this step, and this step can be accomplished effectively using the contact wizard.
6. *Apply the boundary conditions.* Any type boundary conditions can be applied in the contact problems.
7. *Define solution options and load steps.* Convergence behavior for contact problems depends on the geometry and boundaries of the problem. The time step size must be small enough to ensure the convergence. The time step size is specified by the number of steps or the time step size itself. However, the best way to set an accurate time step size is to turn automatic time stepping on in ANSYS solution options.
8. *Solve the contact problem.* Just ensure that the problem is fully converged.
9. *Finally, getting the contact results.* Results from a contact analysis consist mainly of displacements, stresses, strains, reaction forces, and the contact information, such as contact pressure.

Double click on the ANSYS icon

Main Menu > Preferences

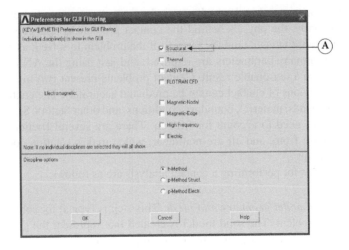

 A select the Structural

OK

This example is limited to structural analysis. Hence, select Structural only. Solid element is used, and its shape is rectangle with four nodes. The contact simulation required another two types of element: the target and contact elements. These two elements will be introduced into the model when the contact wizard tool is used. The following steps are for selecting the element type.

Main Menu > Preprocessor > Element type > Add/Edit/Delete

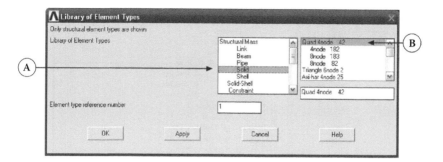

A select Solid

B select Quad 4node 42

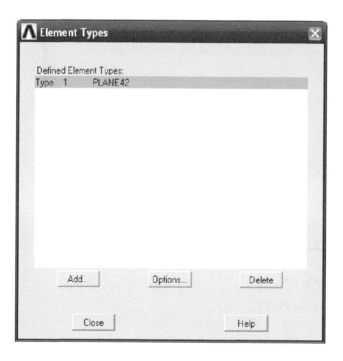

Main Menu > Preprocessor > Material Props > Material Models

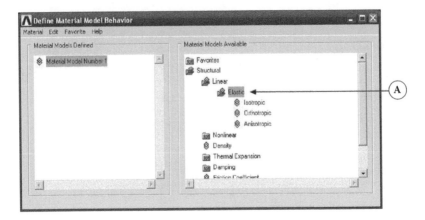

A double click on Structural > Linear > Elastic > Isotropic

The following widows will show up. For the material properties, the modulus of elasticity and Poisson's ratio are required to solve the problem.

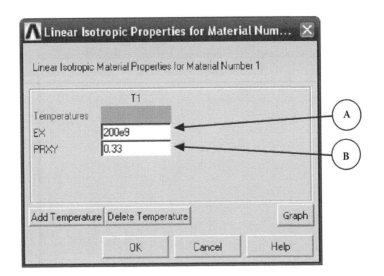

A type 200e9 in EX

B type 0.33 in the PRXY

OK

Close the material model behavior window

Main Menu > Preprocessor > Modeling > Create > Areas > Rectangle > By 2 Corners

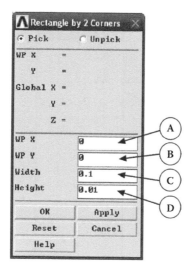

A type 0 in WP X

B type 0 in WP Y

C type 0.1 in Width

D type 0.01 in the Height

Apply

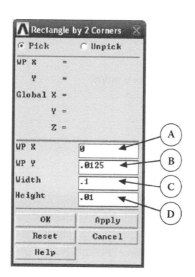

A type 0 in WP X

B type 0.0125 in WP Y

C type 0.1 in Width

D type 0.01 in the Height

OK

Main Menu > Preprocessor > Meshing > Mesh Tool

A select Smart Size

B set Smart Size to 1

C Mesh

Click on Pick All in meshed areas window

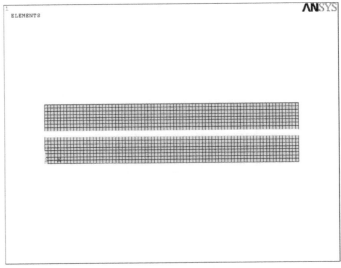

ANSYS graphics shows the mesh

In this problem, the top surface of the lower beam is conventionally designated as the target surface because it is stationary, while bottom surface of the upper beam as the contact surface because it is moving. Both contact and target surfaces are associated with the deformable bodies. These two surfaces together comprise the contact pair. The Contact Manager is very effective in defining, viewing, and editing the contact pairs. In addition, all contact pairs for the entire model can be managed. The Contact Wizard, which is accessed from the Contact Manager, makes the process of creating contact pair very convenient.

Main Menu > Preprocessor > Modeling > Create > Contact Pair

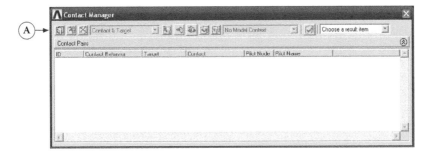

A click on the contact wizard

The contact wizard is used to select the target and contact surfaces. First, the target surface should be selected, then the contact surface.

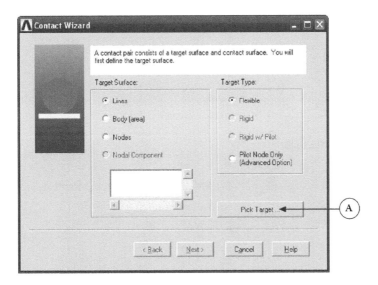

A click on the Pick Target

Using the mouse, click on the top surface of the lower beam.

OK

Next

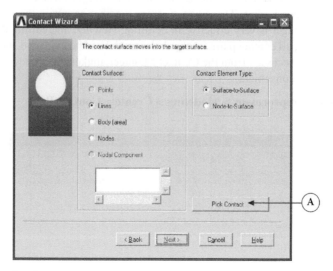

A click on the Pick Contact

Using the mouse, click on the bottom surface of the upper beam.

OK

Next

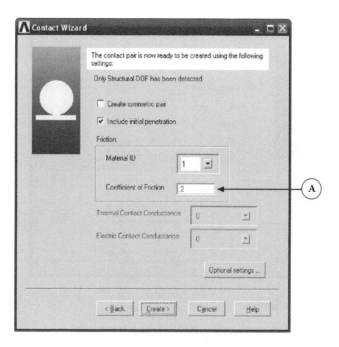

A type 0.2 in the Coefficient of Friction

Create

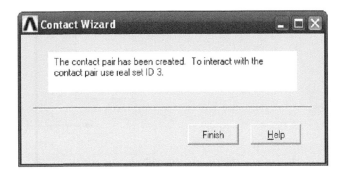

Finish

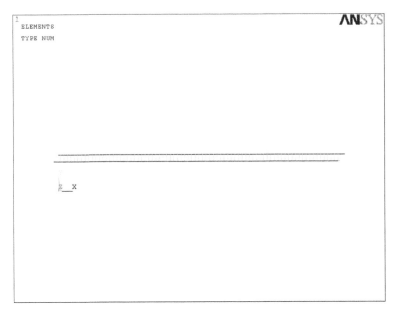

ANSYS graphics shows the contact region

Close the contact manager window

Main Menu > Solution > Define Load > Apply > Structural > Displacement > On Lines

In the ANSYS graphics, click on the vertical right lines of the both beams, where zero *x*- and *y*-displacements are applied.

OK

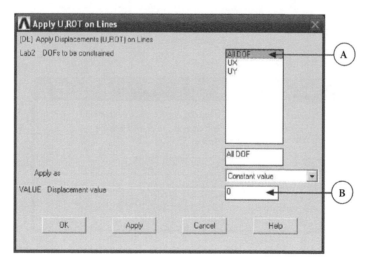

A select All DOF

B type 0 in Displacement value

OK

Pressure is applied at the top surface of the upper beam. A positive pressure means that the pressure is compression.

Main Menu > Solution > Define Load > Apply > Structural > Pressure > On Lines

In the ANSYS graphics, click on the top line of the upper beam.

OK

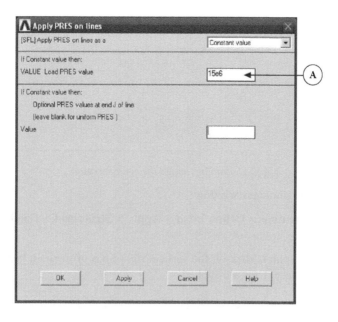

A type 15e6 in Load PRES value

OK

The solution control is quit complicated. The small displacement static is selected. The problem is solved as a static, and the time of the loading process is 1. Using a number higher than 1 will have no effect on the final solution. The total time is divided into 100 subsets. ANSYS is forced to perform at least 50 subsets run regardless of the convergence, and a maximum of 110 subsets runs. In the Frequency, the Write every Nth subsets is selected to ensure that all subsets data are saved for postprocessor.

Main Menu > Solution > Analysis Type > Sol'n Control

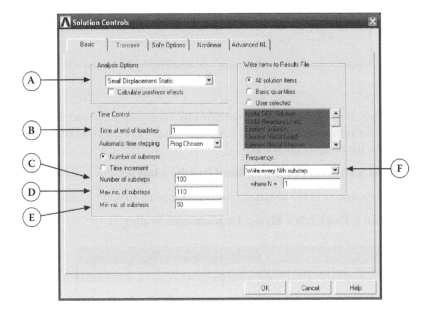

A keep Small Displacement Static

B type 1 in Time end loadstep

C type 100 in Number of steps

D type 110 in Max inc. of substeps

E type 50 in min no. of substep

F select Write every Nth substeps

OK

Main Menu > Solution > Solve > Current LS

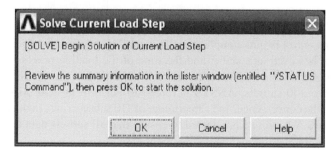

OK

Close

The above windows indicate that the solution task is accomplished successfully. The next step is for getting the results. It is important to upload the last set of the solution iterations for the postprocessor task.

Main Menu > General Postproc > Read Results > last set

To present the results in the actual scaling, the true scale should be selected.

Utility Menu > Plot Ctrls > Style > Displacement Scaling

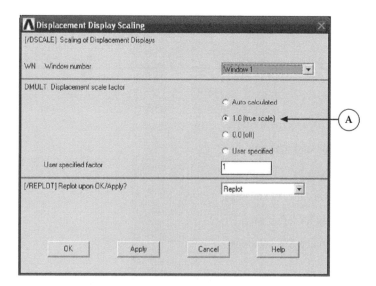

A select 1.0 (true scale)

OK

Main Menu > General Postproc > Plot Results > Deformed Shape

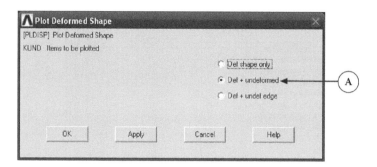

A select Def + undeformed

OK

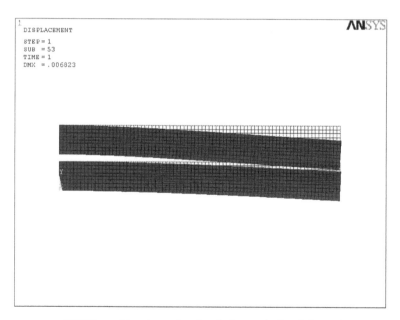

ANSYS graphics shows the plate before and after deformation

The above figure indicates that the upper beam is deflected downward because of the effect of the pressure and reached the lower beam, which is deflected downward. In the next step, the contact status is displayed, maximum displacement and the contact pressure are determined.

Main Menu > General Postproc > Plot Results > Contour Plot > Nodal Solu

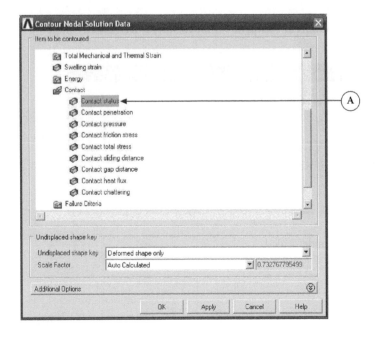

A click on Nodal Solution > Contact > Contact status

OK

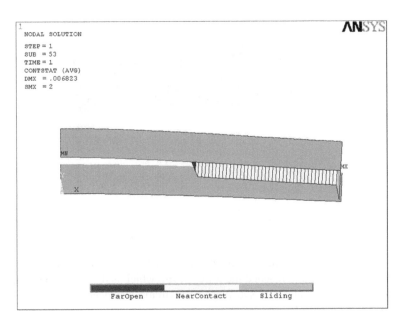

ANSYS graphics shows the location of the contact location

The contact status contours shows the locations of the contact in the problem, including the three regions: no contact (FarOpen), near contact (NearContact), and contact (Sliding).

Main Menu > General Postproc > Plot Results > Contour Plot > Nodal Solu

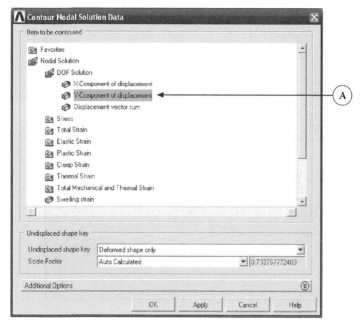

A click on Nodal Solution > DOF Solution > Y-Component of displacement

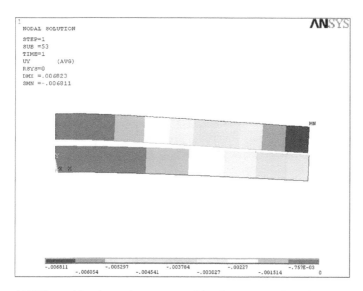

ANSYS graphics shows the contours of displacement in the y-direction

The maximum displacement in the y-direction is occurred in the upper beam, which is equal to 0.00681 m and the maximum displacement for the lower beam is 0.004541 m.

Main Menu > General Postproc > Plot Results > Contour Plot > Nodal Solu

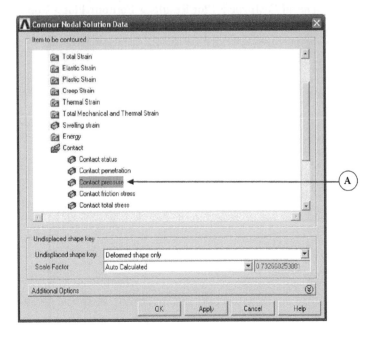

A select Nodal Solution > DOF Solution > Contact pressure

 OK

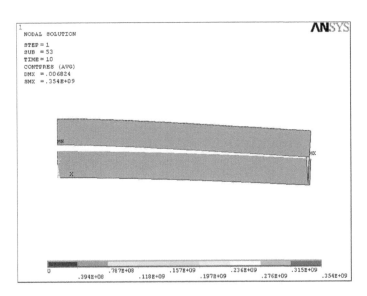

ANSYS graphics shows the contact pressure

The above figure indicates that the maximum contact pressure is equal to 354 MPa.

3.6 VIBRATION ANALYSIS

In this section, the basic formulation of the motion of a single degree of freedom is presented. The aim of this introduction is to discuss some important concepts necessary for solving and understating vibration problems. Figure 3.17 shows a mass–spring system subjected to time-dependent force, $F(t)$, where m is the mass of the system and k is the stiffness of the spring.

Applying the Newton's second law, the equation of motion of the spring–mass element in the x-direction can be expressed as

$$F(t) - kx = m\ddot{x} \tag{3.35}$$

If the applied force, $F(t)$, is equal to zero, Equation 3.35 becomes a homogenous ordinary differential equation. The solution of this equation gives the natural frequency of the oscillating mass

$$\omega^2 = \frac{k}{m} \tag{3.36}$$

where ω is called the natural frequency of the free vibration of the mass. The natural frequency depends only on the spring stiffness and the mass of the system. Substituting Equation 3.36 into the equation of motion of free vibration:

$$\ddot{x} + \omega^2 x = 0 \tag{3.37}$$

In order to derive the finite element equations for the vibration, the element shape is selected. The element, as shown in Figure 3.18, has an initial length L, cross-section area A, and density ρ. Local force are applied at each node.

FIGURE 3.17 Spring–mass system subjected to a time-dependent force.

FIGURE 3.18 Spring element.

The displacement function is assumed linear along the local x-direction as follows:

$$\hat{u} = a_1 + a_2 \hat{x} \tag{3.38}$$

The value of a's can be determined using the boundary conditions and the result is

$$\hat{u} = \left(1 - \frac{\hat{x}}{L}\right)\hat{d}_{1x} + \left(\frac{\hat{x}}{L}\right)\hat{d}_{2x} \tag{3.39}$$

The strain–displacement relationship is given by the equation: $\{\epsilon\} = [B]\{\hat{d}\}$, and $[B]$ and $\{\hat{d}\}$ are

$$[B] = \left[-\frac{1}{L} \quad \frac{1}{L}\right] \tag{3.40}$$

and

$$\{\hat{d}\} = \begin{Bmatrix} \hat{d}_{1x} \\ \hat{d}_{2x} \end{Bmatrix} \tag{3.41}$$

The stress–strain relationship is given by

$$\{\sigma_x\} = [D][\epsilon_x] = [D][B]\{\hat{d}\} \tag{3.42}$$

Applying the Newton's second law at each node, the internal force at each node can be written as

$$\hat{f}_{1x}^e = \hat{f}_{1x} + m_1 \frac{\partial \hat{d}_{1x}}{\partial \hat{t}^2} \tag{3.43}$$

$$\hat{f}_{2x}^e = \hat{f}_{2x} + m_2 \frac{\partial \hat{d}_{2x}}{\partial \hat{t}^2} \tag{3.44}$$

where $\hat{f}^e$ is the applied external force, and m_1 and m_2 are obtained by lumping the total mass of the element, and equally distributed at each node as follows:

$$m_1 = m_2 = \frac{\rho A L}{2} \tag{3.45}$$

The above results can be expressed in a matrix form as follows:

$$
\left\{ \begin{array}{c} \hat{f}_{1x}^{e} \\ \hat{f}_{2x}^{e} \end{array} \right\} = \left\{ \begin{array}{c} \hat{f}_{1x} \\ \hat{f}_{2x} \end{array} \right\} + \left[\begin{array}{cc} m_1 & 0 \\ 0 & m_2 \end{array} \right] \left\{ \begin{array}{c} \dfrac{\partial^2 \hat{d}_{1x}}{\partial \hat{t}^2} \\ \dfrac{\partial^2 \hat{d}_{2x}}{\partial \hat{t}^2} \end{array} \right\}
\tag{3.46}
$$

Using Equation 3.42, and the definition of $\left[\hat{k}\right]$, the matrices (Equation 3.46) can be written as

$$
\{\hat{f}^{e}(t)\} = \left[\hat{k}\right]\{\hat{d}\} + [\hat{m}]\{\hat{\ddot{d}}\}
\tag{3.47}
$$

where

$$
\left[\hat{k}\right] = \frac{AE}{L} \left[\begin{array}{cc} 1 & -1 \\ -1 & 1 \end{array} \right]
\tag{3.48}
$$

$$
[\hat{m}] = \frac{\rho AL}{2} \left[\begin{array}{cc} 1 & 0 \\ 0 & 1 \end{array} \right]
\tag{3.49}
$$

which is also called the lumped-mass matrix, and the acceleration:

$$
\{\hat{\ddot{d}}\} = \frac{\partial^2 \{\hat{d}\}}{\partial \hat{t}^2}
\tag{3.50}
$$

D'Alembert's introduced the effective body force X^e, and body force is in the oppo-site direction of the acceleration, and the body force can be expressed as

$$
\{X^e\} = -\rho\{\ddot{u}\}
\tag{3.51}
$$

Since the element mass matrix can be defined as

$$
[\hat{m}] = \iiint_{V} \rho \left\{ \begin{array}{c} 1 - \dfrac{\hat{x}}{L} \\ \dfrac{\hat{x}}{L} \end{array} \right\} \left[1 - \dfrac{\hat{x}}{L} \quad \dfrac{\hat{x}}{L} \right] dV
\tag{3.52}
$$

The body forces at the node can be evaluated as

$$\{f_b\} = \iiint_V \left\{ \begin{array}{c} 1 - \dfrac{\hat{x}}{L} \\[2mm] \dfrac{\hat{x}}{L} \end{array} \right\} \{X\} dV \tag{3.53}$$

Using the definition of $\{X^e\}$, the body force can be expressed as

$$\{f_b\} = -\iiint_V \rho \left\{ \begin{array}{c} 1 - \dfrac{\hat{x}}{L} \\[2mm] \dfrac{\hat{x}}{L} \end{array} \right\} \{\ddot{u}\} dV \tag{3.54}$$

Since,

$$\{\ddot{u}\} = \left[1 - \dfrac{\hat{x}}{L} \quad \dfrac{\hat{x}}{L} \right] \{\ddot{d}\}$$

the body force can have the following form:

$$\{f_b\} = \iiint_V \rho \left\{ \begin{array}{c} 1 - \dfrac{\hat{x}}{L} \\[2mm] \dfrac{\hat{x}}{L} \end{array} \right\} \left[1 - \dfrac{\hat{x}}{L} \quad \dfrac{\hat{x}}{L} \right] dV \{\ddot{d}\} = -[\hat{m}]\{\ddot{d}\} \tag{3.55}$$

Or, in a simpler form:

$$\{f_b\} = A\rho \int_0^L \rho \left\{ \begin{array}{c} 1 - \dfrac{\hat{x}}{L} \\[2mm] \dfrac{\hat{x}}{L} \end{array} \right\} \left[1 - \dfrac{\hat{x}}{L} \quad \dfrac{\hat{x}}{L} \right] d\hat{x} \tag{3.56}$$

Applying the Newton's second law, the equation of motion of the element in the x-direction can be expressed as

$$\{F(t)\} = [K]\{d\} + [M]\{\ddot{d}\} \tag{3.57}$$

where
  $F(t)$ is the force matrix
  $[K]$ is the stiffness matrix
  $\{d\}$ is the displacement vector
  $[M]$ is the mass matrix
  $\{\ddot{d}\}$ is the acceleration vector

The global matrices are assembled in the same manner as the global stiffness matrix as follows:

$$[F] = \sum_{e=1}^{N} \left[f^{(e)} \right]$$

(3.58)

$$[K] = \sum_{e=1}^{N} \left[K^{(e)} \right]$$

(3.59)

$$[M] = \sum_{e=1}^{N} \left[m^{(e)} \right]$$

(3.60)

where
N is the number of elements
e is the element number

3.7 MODAL VIBRATION FOR A PLATE WITH HOLES

For the geometry shown in Figure 3.19, determine the frequency of first five free vibration modes for a solid beam with six holes. Also, create animation file for the third mode. Let $t = 0.005$ m, $E = 210$ GPa, $v = 0.25$, and $\rho = 5000$ kg/m³.

This example is limited to vibration analysis. Hence, select Structural. The geometry is meshed with triangular six-node elements.

Double click on the ANSYS icon

Main Menu > Preferences

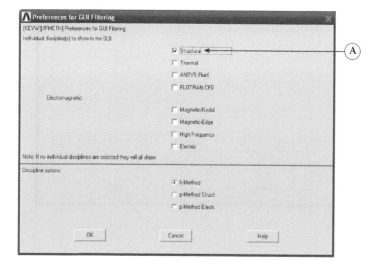

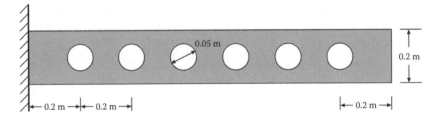

FIGURE 3.19   A plate with six holes.

A select the Structural

OK

Material properties of the structure are defined in following step. It is elastic and independent of the direction, isotropic. The modulus of elasticity, Poisson's ratio, and density are required to solve this problem. In addition, the plate has a thickness, and the thickness is specified in the real constants.

Main Menu > Preprocessor > Element type > Add/Edit/Delete

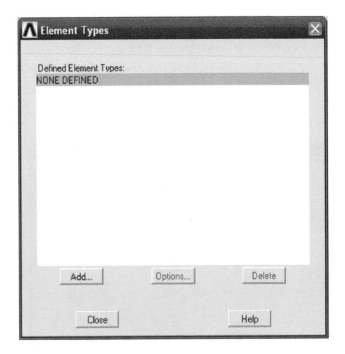

Add...

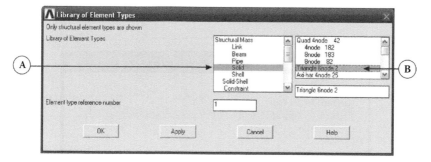

A select Solid

B select Triangle 6node 2

OK

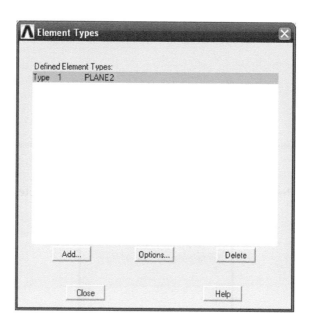

Option

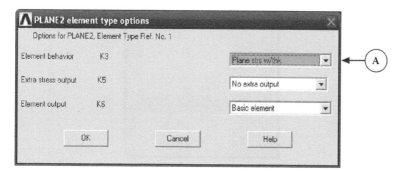

A select Plane strs w/thk

OK

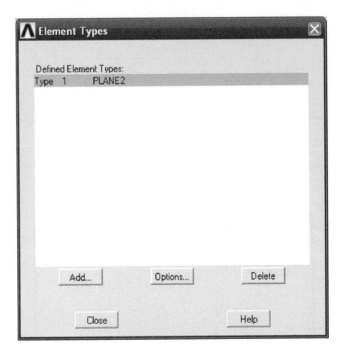

Close

Main Menu > Preprocessor > Real Constants > Add/Edit/Delete

Add

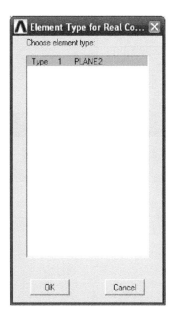

OK

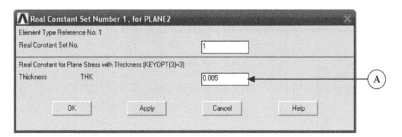

A type 0.005 in Thickness

OK

Close

Main Menu > Preprocessor > Material Probs > Material Models

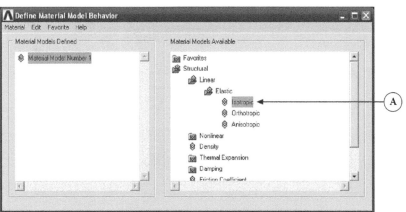

A double click on Structure > Linear > Elastic > Isotropic

The following windows will show up

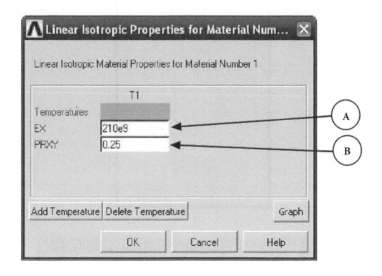

A type 210e9 in EX

B type 0.25 in the PRXY

Main Menu > Preprocessor > Material Props > Material Models

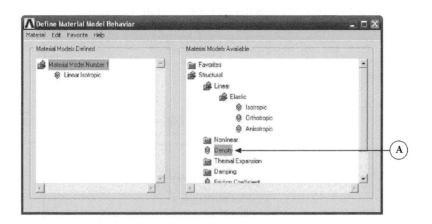

A double click on Structure > Density

The following windows will show up

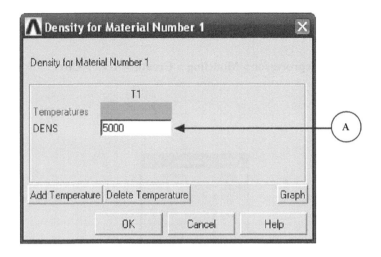

A type 5000 in DENS

 OK

Close the material model behavior window

The geometry is modeled by creating a rectangle and circles. Boolean operation is utilized to remove the circles from the square using overlap and delete commands. Alternatively, the circle can be directly removed by using the subtract command.

Main Menu > Preprocessor > Modeling > Create > Areas > Rectangle > By 2 Corners

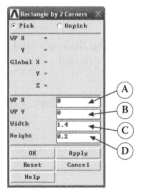

A type 0 in WP X

B type 0 in WP Y

C type 1.4 in Width

D type 0.2 in the Height

OK

Main Menu > Preprocessor > Modeling > Create > Area > Circle > Solid Circle

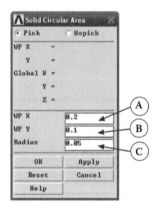

A type 0.2 in WP X

B type 0.1 in WP Y

C type 0.05 in Radius

OK

The other circles can be created using the copy area in the modeling main menu. The number of circles including the original circle is 6 and the space between the circles is equal to 0.2 m.

Main Menu > Preprocessor > Modeling > Copy > Areas

Using the mouse, select the solid circular area

OK

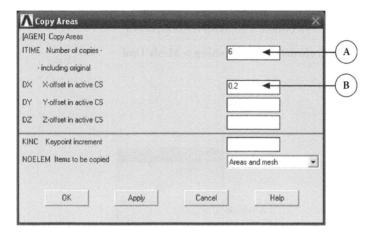

A type 6 in Number of copies

B type 0.2 DX

OK

Main Menu > Preprocessor > Modeling > Operate > Booleans > Overlap > Areas

Click on Pick All in select area to select all areas

Main Menu > Preprocessor > Modeling > Delete > Area Only

Click on the all circles to highlight them

OK

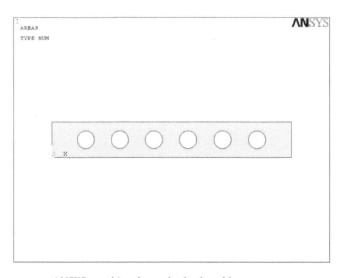

ANSYS graphics shows the final problem geometry

The geometry is meshed with triangular six-node elements. A free mesh is generated using the smart mesh option. The mesh refinement is 1.

Main Menu > Preprocessor > Meshing > Mesh Tool

A select Smart Size

B set Smart Size to 1

C Mesh

click on **Pick All** in mesh area to select the computational domain

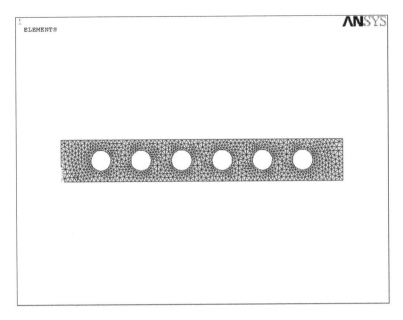

ANSYS graphics shows the mesh

The analysis type is changed from Static to Model. The required number of free modes to be calculated is 5.

Main Menu > Solution > Analysis Type > New Analysis

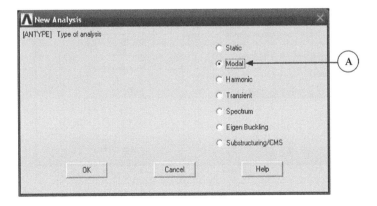

A select Modal

Main Menu > Solution > Analysis Type > Analysis Options

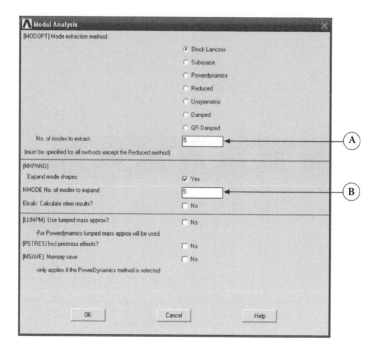

A type 5 in No. of modes to extract

B type 5 in NMODE

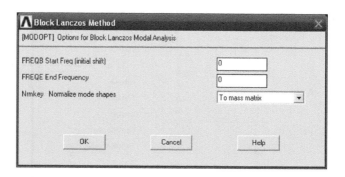

Boundary conditions are applied as follows. The left vertical line of the beam is fixed, while the other lines are free. No force is applied because the case study is free vibration.

Main Menu > Solution > Define Load > Apply > Structural > Displacement > On Lines

In the ANSYS graphics, click on the left line where zero displacement is applied

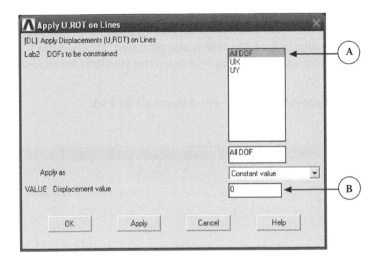

A select All DOF

B set VALUE to 0

OK

Main Menu > Solution > Solve > Current LS

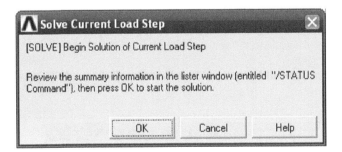

OK

Close

The solution task is done. The solution is ready for the postprocessor analyses. First, the result for a second mode is selected, followed by a display of the static deformation. The deformation of the fifth mode is also presented. Then, a list of frequencies is displayed. Finally, an animation file is created that simulates the vibration for the third mode.

Main Menu > General Postproc > Read Results > By Pick

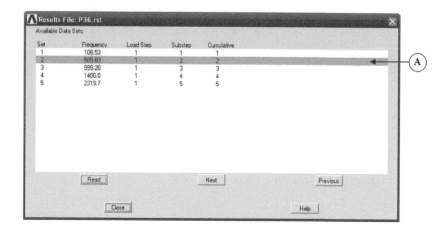

A select Set 2

[Read]

[Close]

Main Menu > General Postproc > Plot Results > Deformed Shape

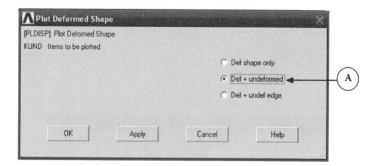

A select Def + undeformed

[OK]

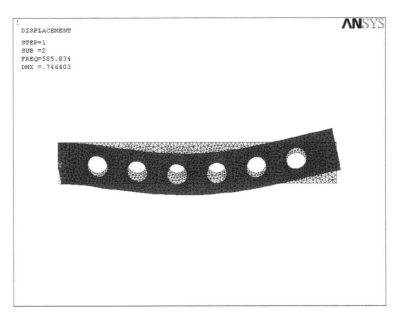

ANSYS graphics shows the beam before and after deformation

Main Menu > General Postproc > Read Results > By Pick

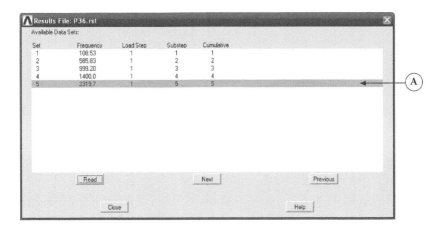

A select Set 5

Read

Close

Main Menu > General Postproc > Plot Results > Deformed Shape

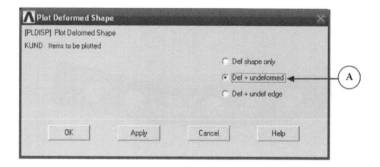

A select Def + undeformed

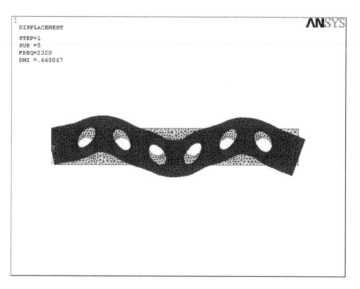

ANSYS graphics shows the beam before and after deformation

To display the frequencies of the vibration mode, a list of results can be used as follows:

Main Menu > General Postproc > List Results > Detailed Summary

```
A  SET,LIST Command                                                    X
File

*****  INDEX OF DATA SETS ON RESULTS FILE  *****
  SET   TIME/FREQ   LOAD STEP   SUBSTEP   CUMULATIVE
   1   108.53          1          1          1

   2   585.83          1          2          2

   3   999.20          1          3          3

   4   1400.0          1          4          4

   5   2319.7          1          5          5
```

The frequencies of the beam without holes can be obtained theoretically, and the first five frequencies are

First mode: f = 106.8306 Hz

Second mode: f = 669.47 Hz

Third mode: f = 1874.5 Hz

Fourth mode: f = 3673.3 Hz

Fifth mode: f = 9070.95 Hz

The first mode, which is 106.83 Hz, is very close to the results obtained by the ANSYS which is 108.53. The difference between the ANSYS and the theoretical solution for the second mode is about 14%. This error is expected because the geometry of the theoretical solution has no holes. The error of the third, fourth, and fifth frequencies is large. Next, an animation file is created for the third vibration mode.

Main Menu > General Postproc > Read Results > By Pick

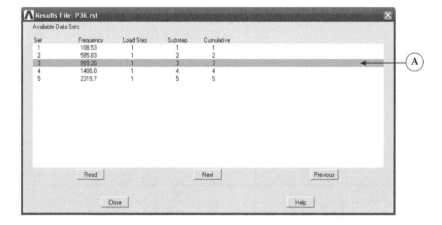

A select Set 3

Read

Close

Main Menu > General Postproc > Plot Results > Deformed Shape

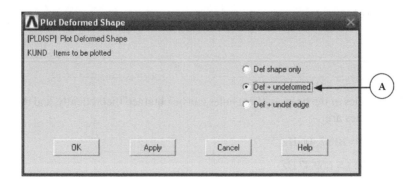

A select Def + undeformed

OK

Utility Menu > PlotCtrls > Animate > Mode Shape

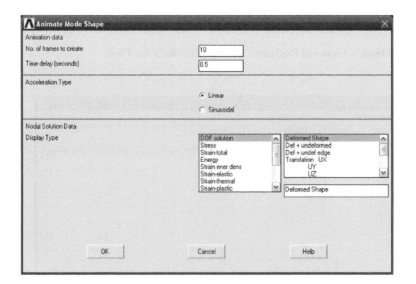

OK

Animation will be shown in the screen of the third mode. The number of frames to create in the above window is for the number of screen shoots for the animation file, and a higher value will generate a better resolution movie, but with a larger file size. The time delay is the time between each screen shots, and increasing the time delay will increase the duration of the movie. The total duration of the move is simply the number of frames times the time delay, which is 5 s.

3.8 HARMONIC VIBRATION FOR A PLATE WITH HOLES

For the plate with holes shown in Figure 3.20, a load is applied at the end of the plate as shown in the figure. The magnitude of the force is 500 N, and the frequency of the load is varied between 1 and 700 Hz. Let $t = 0.005$ m, $E = 210$ GPa, $v = 0.25$, and $\rho = 5000$ kg/m³. Create a graph showing the relationship between the displacement at the point where the force is applied and the load's frequency.

Since the geometry of the present problem is the same as in the previous example, follow the steps in the previous example to model and mesh the geometry.

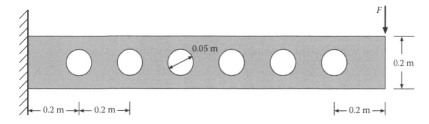

FIGURE 3.20 A plate with six holes.

Main Menu > Solution > Analysis Type > New Analysis

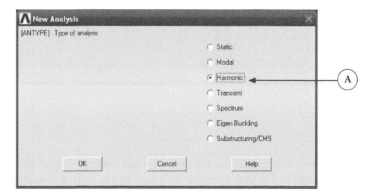

A select Harmonic

 OK

This example illustrates the full method for the harmonic vibration because it is simple compared to the reduced method. However, the full method makes use the full stiffness and mass matrices and thus slower and costlier option.

Main Menu > Solution > Analysis Type > Analysis Options

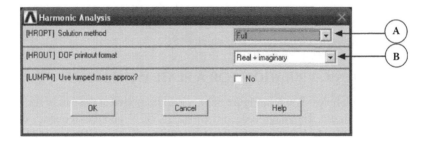

A select Full

B select Real + imaginary

The following window will show up. Use the default settings

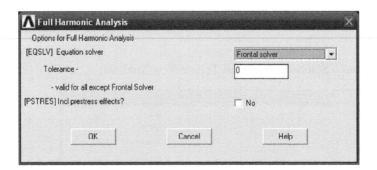

Boundary conditions are applied as follows. The left vertical line of the beam is fixed, while the other end is free. Force in the negative y-direction is applied on the node at the upper right corner.

Main Menu > Solution > Define Load > Apply > Structural > Displacement > On Lines

In the ANSYS graphics, click on the left line where displacement is applied

OK

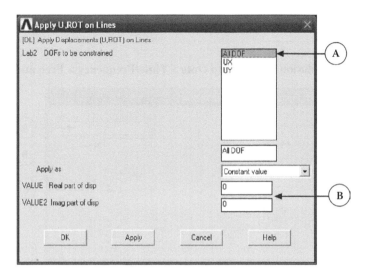

A select All DOF

B set VALUE and VALUE 2 to 0

 OK

Next, the information about the magnitude and phase of the load are provided to the ANSYS solver. The magnitude of the load is 500 N and its phase is 0. The phase is important when two or more cyclic loads are applied. However, for a harmonic analysis, all loads must have the same frequencies.

Main Menu > Solution > Define Load > Apply > Structural > Force/Moment > On Node

In the ANSYS graphics, click on the node where force is applied

 OK

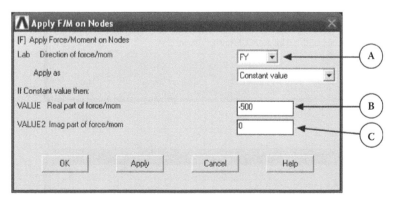

A select FY

B type -500 in Real part of force/mom

C type 0 in Imag of force/mom

Main Menu > Solution > Load Setp Opts > Time/Frequency > Freq and Substps

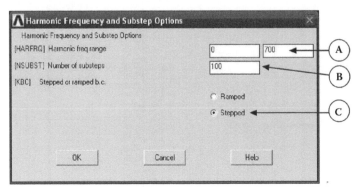

A type 0 and 700 in the Harmonic freq range

B type 100 in Number of substeps

C select Stepped

OK

The plate will be subjected to a force at 1, 2, 3, …, 700 Hz. By specifying the stepped boundary load, the load of 500 N will be applied at each of the frequencies. If the ramped option is used, the magnitude of the load will be 1 N at 1 Hz, 250 N at 350 Hz, and 500 N at 700 Hz.

Main Menu > Solution > Solve > Current LS

OK

$\boxed{\text{OK}}$

A graph that shows the displacement at a specific location in the plate as a function of frequency can be done easy with the time history feature of the ANSYS. In the following steps, the displacement at the node where force is applied is plotted as a function of frequency.

Main Menu > TimeHist Postpro

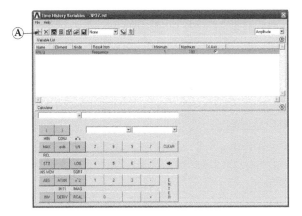

A click on the green X button

Using the mouse, click on the node at the upper right corner of the plate in the ANSYS graphics area.

$\boxed{\text{OK}}$

The following windows will show up

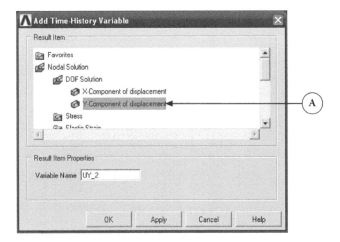

A click on Nodal Solution > DOF Solution > Y-Component of displacement

In the ANSYS graphics, click on the node where force is applied

OK

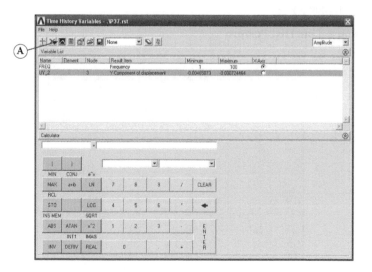

A click on the block graph button

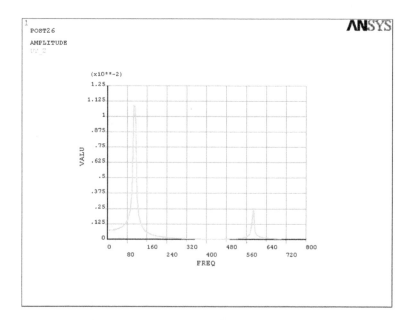

ANSYS graphics y-displacement as a function of frequency

The results from the displacement versus frequency are consistent with the result obtained in the previous problem. The jumps in the displacement shown in the above figure are occurred at the natural frequencies of the first and second modes of the plate, which are 108.53 and 585.83 Hz. To observe the other natural frequencies, the frequency range should be increased from 700 to 2400 Hz to have all five natural frequencies.

3.9 HIGHER-ORDER ELEMENTS

In this section, the stiffness matrix for higher-order elements is presented. The first higher-order type is a triangular element with six nodes. Each node can carry 2 degrees of freedom and total of 12 degrees of freedom for each element. Figure 3.21 shows the basic six-node triangular element. The procedures for development of the equations are similar to the procedures for a linear triangular element. Since there are 6 degrees of freedom for the x-displacement and 6 more for the y-displacement, the displacement functions should be quadratic:

$$u(x, y) = a_1 + a_2 x + a_3 y + a_4 x^2 + a_5 xy + a_6 y^2 \tag{3.61}$$

$$v(x, y) = a_7 + a_8 x + a_9 y + a_{10} x^2 + a_{11} xy + a_{12} y^2 \tag{3.62}$$

The unknown nodal displacements can be expressed in a vector form as follows:

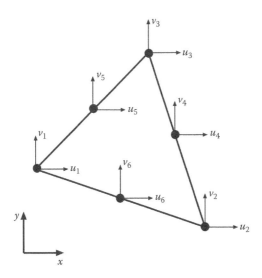

FIGURE 3.21 Six-node triangular element.

$$\{\underline{d}\} = \begin{Bmatrix} \underline{d_1} \\ \underline{d_2} \\ \underline{d_3} \\ \underline{d_4} \\ \underline{d_5} \\ \underline{d_6} \end{Bmatrix} = \begin{Bmatrix} u_1 \\ v_1 \\ u_2 \\ v_2 \\ u_3 \\ v_3 \\ u_4 \\ v_4 \\ u_5 \\ v_5 \\ u_6 \\ v_6 \end{Bmatrix} \qquad (3.63)$$

The stress–strain matrix for six-node triangular elements is same as for the six-node triangular elements. [D] matrix is defined as

$$[D] = \frac{E}{1-v^2} \begin{bmatrix} 1 & v & 0 \\ v & 1 & 0 \\ 0 & 0 & \dfrac{1-v}{2} \end{bmatrix} \qquad (3.64)$$

The [B] matrix is a function of x and y:

$$[B] = \frac{1}{2A} \begin{bmatrix} \beta_1 & 0 & \beta_2 & 0 & \beta_3 & 0 & \beta_4 & 0 & \beta_5 & 0 & \beta_6 & 0 \\ 0 & \gamma_1 & 0 & \gamma_2 & 0 & \gamma_3 & 0 & \gamma_4 & 0 & \gamma_5 & 0 & \gamma_6 \\ \gamma_1 & \beta_1 & \gamma_2 & \beta_2 & \gamma_3 & \beta_3 & \gamma_4 & \beta_4 & \gamma_5 & \beta_5 & \gamma_6 & \beta_6 \end{bmatrix} \qquad (3.65)$$

where

$$\beta_1 = -3h + \frac{4hx}{b} + 4y \qquad \beta_2 = -h + \frac{4hx}{b} \qquad \beta_3 = 0$$

$$\beta_4 = 4y \qquad \beta_5 = -4y \qquad \beta_6 = 4h - \frac{8hx}{b} - 4y$$

$$\gamma_1 = -3b + 4x + \frac{4by}{h} \qquad \gamma_2 = 0 \qquad \gamma_3 = -b + \frac{4by}{h}$$

$$\gamma_4 = 4x \qquad \gamma_5 = 4b - 4x - \frac{8by}{h} \qquad \gamma_6 = -4x$$

b and h are the width and the height of the element, respectively. The stiffness matrix is determined by performing the integration of $[k]$:

$$[K] = \iiint [B]^T [D][B] dV \qquad (3.66)$$

The stress can also be determined using the following equation:

$$\{\sigma\} = [d]\{\epsilon\} \qquad (3.67)$$

ANSYS software called the six-node triangular element PLANE2 element. This element has a quadratic displacement behavior, and it is well suited to model irregular geometries.

Rectangular elements are also commonly used to mesh solid structures. A linear rectangular element, Figure 3.22, has four nodes at the corners. Each node has 2 degrees of freedom. ANSYS has this type of element in its library, and it is called PLANE42. The displacement function for this element is

$$u(x, y) = a_1 + a_2 x + a_3 y + a_4 xy \qquad (3.68)$$

$$v(x, y) = a_5 + a_6 x + a_7 y + a_8 xy \qquad (3.69)$$

A higher-order rectangular element is the quadratic element. This element has eight nodes, four at the corners and four at the sides, as shown in Figure 3.23. With this element, the sides can be bended to model complex geometries. ANSYS has this type of element in its library, and it is called PLANE82. The displacement function is given by

$$u(x, y) = a_1 + a_2 x + a_3 y + a_4 x^2 + a_5 xy + a_6 y^2 + a_7 x^2 y + a_8 xy^2 \qquad (3.61)$$

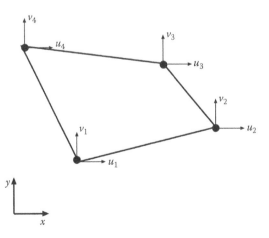

FIGURE 3.22 Four-node rectangular element.

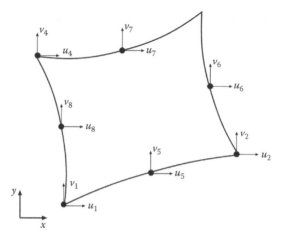

FIGURE 3.23   Eight-node rectangular element.

$$v(x, y) = a_9 + a_{10}x + a_{11}y + a_{12}x^2 + a_{13}xy + a_{14}y^2 + a_{15}x^2y + a_{16}xy^2 \qquad (3.62)$$

PROBLEMS

3.1 The plate shown in Figure 3.24 is subjected to a compression pressure, and the applied pressure is 25 MN/m². Let $E = 210$ GPa, $v = 0.25$, and $t = 0.005$ m. Use ANSYS to determine the maximum stress in the x- and y-directions, and also the maximum displacement in the y-direction. Use triangular six-node elements with smart size mesh of 1.

3.2 A solid plate is loaded with 250 MN/m² on both sides as shown in Figure 3.25. Obtain the maximum stresses in the x-direction. Also, obtain the stress concentration factor (K). Compare the ANSYS result with maximum stress using concentration factor using Figure 3.11. Given $E = 210$ GPa and $v = 0.3$. Use triangular six-node elements with smart size mesh of 1.

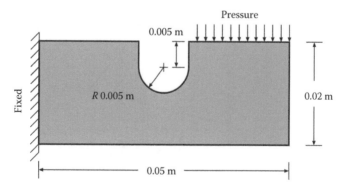

FIGURE 3.24   A plate subjected to a compression pressure for Problem 3.1.

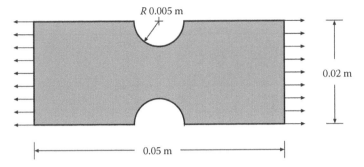

FIGURE 3.25 A solid plate loaded on both sides for Problem 3.2.

3.3 A solid plate with four holes is subjected to pressure of $100\,MN/m^2$ on both sides as shown in Figure 3.26. Obtain the maximum stress and displacement in the x-direction. Given $E = 190\,GPa$, $v = 0.33$, and $t = 0.001\,m$. Use triangular six-node elements with smart size mesh of 1.

3.4 A solid mechanical member is loaded as shown in Figure 3.27. Determine the magnitude and direction of the maximum displacements. Solve the problem in three-dimensional space. The applied force is $100\,N$. Given $E = 210\,GPa$ and $v = 0.25$. Use Tel 10node element with smart size of 10.

3.5 For the plate shown in Figure 3.28, determine the first five vibration modes of natural frequency. Also, create animation file for the third vibration mode. Let $t = 0.005\,m$, $E = 180\,GPa$, $v = 0.33$, and $\rho = 4500\,kg/m^3$. Use quadratic four-node elements with smart size mesh of 1.

3.6 For the geometry shown in Figure 3.29, a load is applied at the end of the plate, as shown in Figure 3.29. Let $t = 0.005\,m$, $E = 180\,GPa$, $v = 0.33$, and $\rho = 4500\,kg/m^3$. Create a graph showing the relationship between the displacement in the x- and y-directions at the point where the force is applied and the load frequency. The

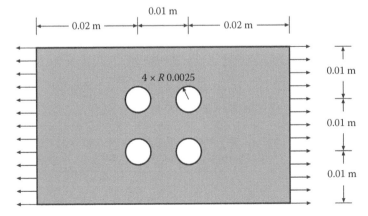

FIGURE 3.26 A solid plate with four holes for Problem 3.3.

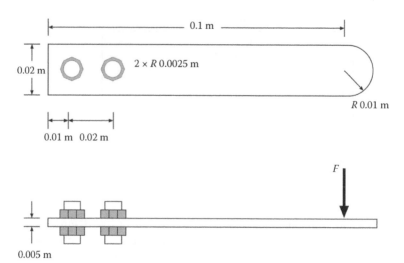

FIGURE 3.27   A solid mechanical member for Problem 3.4.

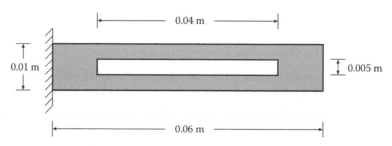

FIGURE 3.28   Geometry for Problem 3.5.

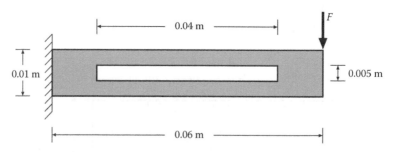

FIGURE 3.29   Geometry for Problem 3.6.

magnitude of the force is 100 N, and the frequency of the load is varied between 1 and 10,000 Hz with 200 subsets. Use quadratic four-node elements with smart size mesh of 1.

3.7 For the airplane wing shown in Figure 3.30, determine the first five modes of natural frequency. Let $E = 180$ GPa, $v = 0.25$, and $\rho = 3500$ kg/m^3. Solve it as three-dimensional problem. The left end of the wing is fixed. Use tetrahedral 10-node elements with smart size mesh of 1.

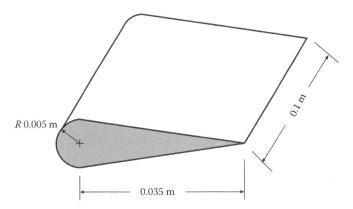

FIGURE 3.30 Airplane wing for Problem 3.7.

3.8 A solid mechanical member is loaded as shown in Figure 3.31. Determine the first three modes of the natural frequency. Given $E = 210\,\text{GPa}$, $v = 0.25$, and $\rho = 3200\,\text{kg/m}^3$. Use tetrahedral 10-node elements with smart size mesh of 5.

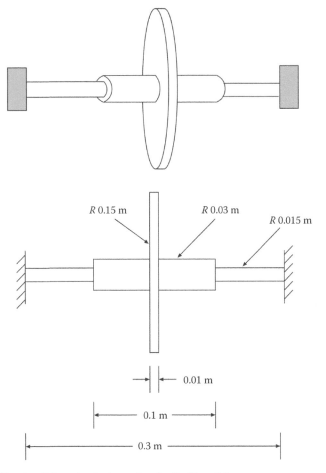

FIGURE 3.31 A solid mechanical member for Problem 3.8.

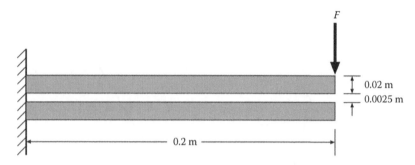

FIGURE 3.32   A contact element problem.

3.9 Two horizontal beams are placed close to each other, as shown in Figure 3.32. Both beams are fixed at the left side, and force of 200 kN is applied to the upper beam. As a result, the upper beam will deflect and meet the lower beam. Determine the maximum deflection of the first and second beams in the y-direction, and contact pressure. Given $E = 180\,\text{GPa}$ and $v = 0.25$ for both beams. The friction coefficient between the two beams is estimated to be 0.1.

4 Heat Transfer

4.1 INTRODUCTION TO HEAT CONDUCTION

Whenever a temperature gradient exits in a solid, heat will flow from a high-temperature region to a low-temperature region. The basic governing heat conduction equations are obtained by considering a plate with a surface area A and a thickness Δx, as shown in Figure 4.1. One side is maintained at a temperature T_1, and the other side at temperature T_2. Experimental observation indicates that the rate of heat flow is directly proportional to the area and temperature difference, but inversely proportional to the plate thickness. The proportionality sign is replaced by an equal sign by introducing the constant k, as follows

$$Q = kA\frac{T_1 - T_2}{\Delta x} \tag{4.1}$$

where k is the thermal conductivity of the plate, and this property depends on the type of the plate material. Equation 4.1 is also called Fourier's law. Fourier's law can also be expressed in differential form in the direction of the normal coordinate:

$$Q = -kA\frac{dT}{dn} \tag{4.2}$$

Also, Fourier's law can be expressed for multidimensional heat flux flow:

$$Q'' = -k\left(\frac{\partial T}{\partial x}i + \frac{\partial T}{\partial y}j + \frac{\partial T}{\partial z}k\right) \tag{4.3}$$

An energy balance can be applied to a differential volume, $dx\,dy\,dz$, for conduction analysis in a Cartesian coordinate, as shown in Figure 4.2. The objective of this energy balance is to obtain the temperature distribution within the solid. The temperature distribution can be used to determine the heat flow at a certain surface, or to study the thermal stress.

The heat flux perpendicular to the surface of the control volume is indicated by the terms, Q_x'', Q_y'', and Q_z''. The heat flux at the opposite surfaces can be then expressed using the Taylor series expansion of the first order, as follows:

$$Q_{x+dx}'' = Q_x'' + \frac{\partial Q_x''}{\partial x}dx \tag{4.4}$$

183

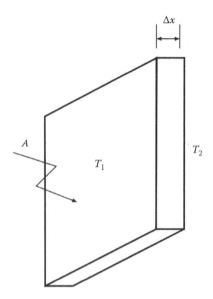

FIGURE 4.1   Heat transfer through a plate.

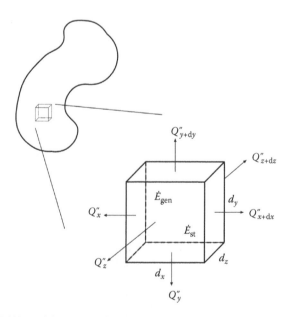

FIGURE 4.2   Differential control volume for the energy balance.

$$Q''_{y+dy} = Q''_y + \frac{\partial Q''_y}{\partial y} dy \tag{4.5}$$

$$Q''_{z+dz} = Q''_z + \frac{\partial Q''_z}{\partial z} dz \tag{4.6}$$

Energy can be generated in the medium. The expression of the heat generation is

$$\dot{E}_{gen} = \dot{q}\, dx\, dy\, dz \tag{4.7}$$

where $\dot{q}$ is the generated heat per unit volume, W/m³. If the heating process is unsteady, the total energy of the control volume can be increased or decreased. The energy storage term is expressed as

$$\dot{E}_{st} = \rho C_p \frac{\partial T}{\partial t} dx\, dy\, dz \tag{4.8}$$

The sum of the energy generation in the control volume, and net heat flow should be equal to the energy stored in the control volume. The energy conservation can be expressed in the following mathematical form:

$$\dot{E}_{gen} + \left(\dot{E}_{in} - \dot{E}_{out} \right) = \dot{E}_{st} \tag{4.9}$$

Substituting expressions (Equations 4.4 through 4.8 into Equation 4.9), the energy conservation equation becomes

$$\dot{q}\, dx\, dy\, dz + \left(\frac{\partial Q''_x}{\partial x} dy\, dz + \frac{\partial Q''_y}{\partial y} dx\, dz + \frac{\partial Q''_z}{\partial z} dx\, dy \right) = \rho C_p \frac{\partial T}{\partial t} dx\, dy\, dz \tag{4.10}$$

The Q''_x, Q''_y, and Q''_z can be obtained from Fourier's law (Equation 4.2), as follows:

$$Q''_x = -k \frac{dT}{dx} dx \tag{4.11}$$

$$Q''_y = -k \frac{dT}{dy} dy \tag{4.12}$$

$$Q''_z = -k \frac{dT}{dx} dz \tag{4.13}$$

Finally, the conduction energy equation per unit volume, in the Cartesian coordinate, can be expressed, as follows:

$$\frac{\partial}{\partial x}\left(k\frac{\partial T}{\partial x}\right)+\frac{\partial}{\partial y}\left(k\frac{\partial T}{\partial y}\right)+\frac{\partial}{\partial z}\left(k\frac{\partial T}{\partial z}\right)+\dot{q}=\rho C_p\frac{\partial T}{\partial t}$$ (4.14)

When the system reaches a steady-state condition, the term $\partial T/\partial t$ is eliminated. If the thermal conductivity is independent of the direction, the conduction energy equation can be written in a simpler form:

$$\frac{\partial^2 T}{\partial x^2}+\frac{\partial^2 T}{\partial y^2}+\frac{\partial^2 T}{\partial z^2}+\frac{\dot{q}}{k}=\frac{\rho C_p}{k}\frac{\partial T}{\partial t}$$ (4.15)

The energy equation is a partial differential equation with second order in space and first order in time. The boundary conditions along its surface, as well as the initial condition must be specified. For the initial condition, the temperature distribution in the system must be provided. In heat transfer problems, there are three types of boundary conditions: temperature, heat flux, and convection.

The constant temperature, also called the Dirichlet condition, corresponds to a situation for which the surface is maintained at a fixed temperature all the time. The mathematical expression for this boundary condition is as follows:

$$T(x,t)=T_s$$ (4.16)

The second boundary condition, also called the Neumann condition, corresponds to a constant heat flux applied to a surface. The heat flux is related to the temperature gradient at the surface by Fourier's law, as follows:

$$-k\frac{\partial T}{\partial x}=q''_s$$ (4.17)

A special case of the Neumann boundary condition is an insulated boundary condition, and the heat flux should be zero:

$$-k\frac{\partial T}{\partial x}=0$$ (4.18)

The third boundary condition corresponds to convection at a surface. The conduction–convection heat balance at the wall surface must be satisfied. The heat transfer coefficient (h) should be known, as well as the fluid bulk temperature (T_∞):

$$-k\frac{\partial T}{\partial x}=h[T_\infty-T(x,t)]$$ (4.19)

4.2 FINITE ELEMENT METHOD FOR HEAT TRANSFER

The finite element method is an efficient way to solve conduction problems. The heat transfer solution can be used to estimate the heat flow at the system's boundary, or to determine the temperature distribution for thermal-stress analysis. The first step in the finite element formulation for conduction heat transfer is to select the element type. A linear triangular element is selected because it is the simplest form of elements for two-dimensional analysis. The temperature at the nodes, T_i, T_j, and T_m, are expressed in the following matrix form

$$\{\underline{T}\} = [N_i \quad N_j \quad N_m] \begin{Bmatrix} T_i \\ T_j \\ T_m \end{Bmatrix} \tag{4.20}$$

where Ns are linear shape functions given by

$$N_i = \frac{1}{2A}(\alpha_i + \beta_i x + \gamma_i y) \tag{4.21}$$

$$N_j = \frac{1}{2A}(\alpha_j + \beta_j x + \gamma_j y) \tag{4.22}$$

$$N_m = \frac{1}{2A}(\alpha_m + \beta_m x + \gamma_m y) \tag{4.23}$$

The expression for α's, β's, and γ's are defined as follows:

$$\begin{aligned}
\alpha_i &= x_j y_m - y_j x_m & \alpha_j &= x_m y_i - y_m x_i & \alpha_m &= x_i y_j - y_i x_j \\
\beta_i &= y_j - y_m & \beta_j &= y_m - y_i & \beta_m &= y_i - y_j \\
\gamma_i &= x_m - x_j & \gamma_j &= x_i - x_m & \gamma_m &= x_j - x_i
\end{aligned} \tag{4.24}$$

The temperature gradient matrix is equivalent to the strain matrix used in the stress analysis problems:

$$\{\underline{g}\} = \begin{Bmatrix} \dfrac{\partial T}{\partial x} \\ \dfrac{\partial T}{\partial y} \end{Bmatrix} \tag{4.25}$$

The heat flux and temperature gradient are related to each other using the thermal conductivity matrix [D] as follows

$$\begin{Bmatrix} g_x \\ g_y \end{Bmatrix} = -[D]\{\underline{g}\} \tag{4.26}$$

and the thermal conductivity matrix $[D]$ is defined as

$$[D] = \begin{bmatrix} K_{xx} & 0 \\ 0 & K_{yy} \end{bmatrix} \tag{4.27}$$

Using Equation 4.20 in Equation 4.25, we have

$$\{g\} = \begin{bmatrix} \dfrac{\partial N_i}{\partial x} & \dfrac{\partial N_j}{\partial x} & \dfrac{\partial N_m}{\partial x} \\[2mm] \dfrac{\partial N_i}{\partial y} & \dfrac{\partial N_j}{\partial y} & \dfrac{\partial N_m}{\partial y} \end{bmatrix} \begin{Bmatrix} T_i \\ T_j \\ T_m \end{Bmatrix} \tag{4.28}$$

The temperature gradient matrix can also be written in a compact form:

$$\{g\} = [\underline{B}]\{T\} \tag{4.29}$$

The $[B]$ matrix is defined as

$$[\underline{B}] = \frac{1}{2A} \begin{bmatrix} \beta_i & \beta_j & \beta_m \\ \gamma_i & \gamma_j & \gamma_m \end{bmatrix} \tag{4.30}$$

The stiffness matrix is obtained from the potential energy theory as follows

$$[K] = \iiint_V [\underline{B}]^T [D][\underline{B}]dV + \iint h[N]^T[N]dS \tag{4.31}$$

where the first term contributes for the conduction, while the second term contributes for convection. The element equation should be formulated in the form of $\{f\} = [K]\{T\}$ and the force matrix represents heat flow at the element's boundary, and it is defined as

$$\{f\} = \frac{QAL + q''PL + hT_\infty PL}{2} \begin{Bmatrix} 1 \\ 1 \end{Bmatrix} \tag{4.32}$$

where
  P is the perimeter of the element
  A is the area perpendicular to heat flow
  Q is the heat generation in the element
  q'' is the heat flux at the boundary of the element
  L is the element's side length

4.3 THERMAL ANALYSIS OF A FIN AND A CHIP

Examine the performance of a straight fin used for cooling an electronic chip by determining the maximum operating temperature of the chip at the steady-state condition.

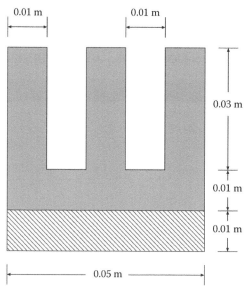

FIGURE 4.3   Chip and fin.

Also, determine the average temperature at the base of the chip. The fin is made of pure aluminum, while the chip is made of epoxy. Free convection boundary condition is imposed at the fin surface and the vertical sides of the chip with $h = 5\,\text{W/m}^2\text{-°C}$ and $T_0 = 25\text{°C}$, while the chip's bottom surface is well insulated. A power of 15 W is generated in the chip. Let $k_{chip} = 0.2\,\text{W/m-°C}$ and $k_{Al} = 237\,\text{W/m-°C}$ (Figure 4.3).

This example is limited to thermal analysis. Hence, select Thermal in the preferences. The Solid element is used, and its shape is a triangle with six nodes.

Double click on the ANSYS icon

Main Menu > Preferences

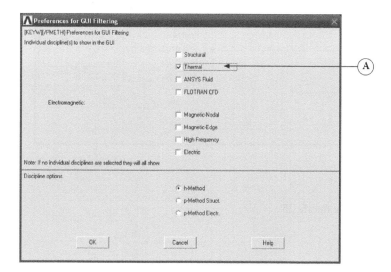

A select the Thermal

 OK

Main Menu > Preprocessor > Element type > Add/Edit/Delete

Add...

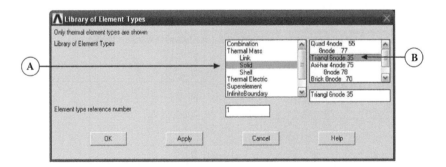

A select Solid

B select Triangle 6node 35

 OK

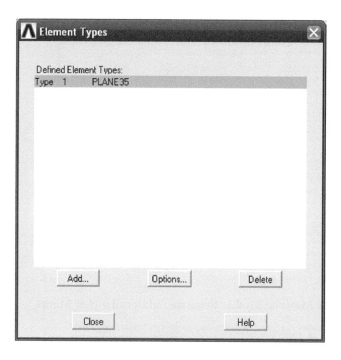

Close

Only thermal conductivity is required to solve the problem. Notice that the chip has a different thermal conductivity than the fin. By default, all areas will be assigned to material number 1. In this problem, the material number of the fin is 1, while the material number of the chip is 2.

Main Menu > Preprocessor > Material Props > Material Models

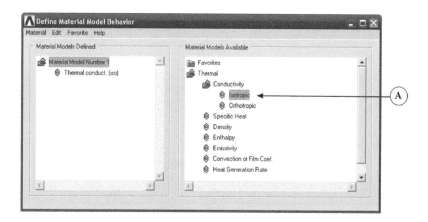

A double click on Thermal > Conductivity > Isotropic

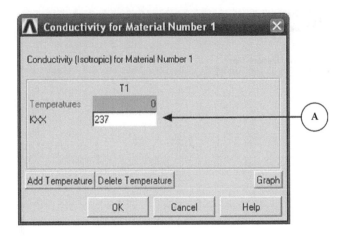

A type 237 in KXX

OK

In the Define Material Models Behavior: Material > New Model

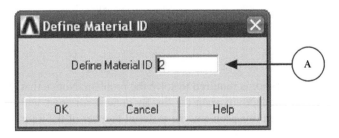

A type 2 in Define Material ID

OK

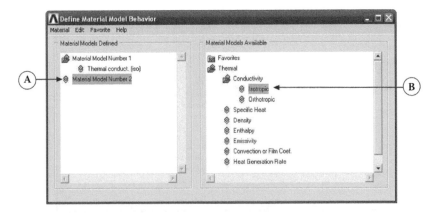

A select Material Model Number 2

B double click on Thermal > Conductivity > Isotropic

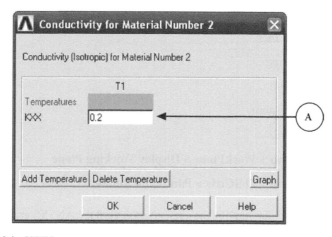

A type 0.2 in KXX

OK

Close the material model behavior window

The geometry is created using the ANSYS graphics. Setting up the work space is done using the WP Setting. Snap is enabled to allow mouse-click on the ANSYS graphics with an increment. Spacing is the distance between the vertical or horizontal grids. The size of the grid is specified in the Minimum and Maximum. The space is divided into squares with side length of 0.01 m. The total width and height of the grids is 0.05 m. This setup makes the modeling easy by creating key points, lines, and areas on the ANSYS graphics.

Utility Menu > WorkPlane > WP Setting

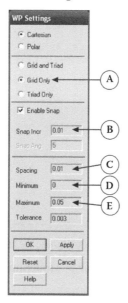

A select Grid only

B type 0.01 in Snap Incr

C type 0.01 in Spacing

D type 0 in Minimum

E type 0.05 in Maximum

OK

ANSYS Utility Menu > WorkPlane > Display Working Plane

ANSYS Utility Menu > PlotCtrls > Pan Zoom Rotate...

Click on zoom in and out, moving the cursor button until the ANSYS graphics shows all grids.

The key points are created first, and then the lines between two key points. Finally, fin and chip areas are created. The thermal conductivity of the chip and fin are assigned using element attribute in the meshing tools.

ANSYS Main Menu > Preprocessor > Modeling > Create > Key points > On Working Plane

By using the mouse, click on the ANSYS graphics window at the location Key points, as shown in the figure below.

OK

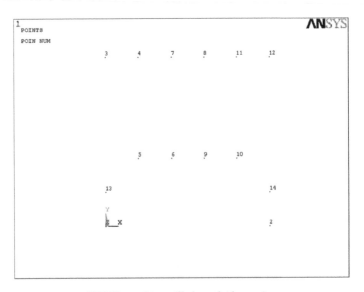

ANSYS graphics will show the key points

ANSYS Main Menu > Preprocessor > Modeling > Create > Lines > Lines > Straight Line

Click on the two Key points to create one straight line, and redo for all lines. The created lines are as shown below.

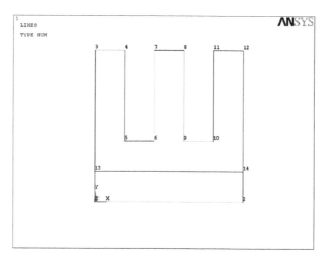

ANSYS graphics shows lines

ANSYS Main Menu > Preprocessor > Modeling > Create > Areas > Arbitrary > By Lines

Click on the chip lines to create a rectangle chip area then , redo to create the fin area. The final geometry is shown below.

OK

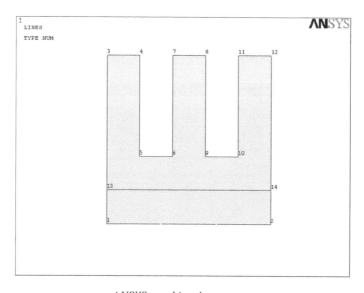

ANSYS graphics shows areas

Main Menu > Preprocessor > Meshing > Mesh Tool

A select Areas

B click on Set

Using mouse, select the chip area only.

OK

The following windows will show up. By selecting number 2, the number 2 properties in the material model are assigned to the chip. The fin, by default, has the properties of number one in the material model.

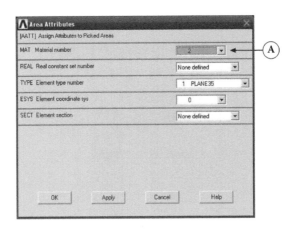

A select 2 in Material number

To ensure that thermal conductivity of the fin and chip are assigned correctly, the chip and fin are colored according to their material number. This step has no effect on the solution. The geometry is meshed with triangular six-node elements. A free mesh is generated using the smart mesh option. The mesh refinement is 1.

Utility Menu > PlotCtrls > Numbering...

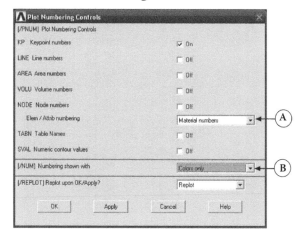

A select Material number

B select Colors only

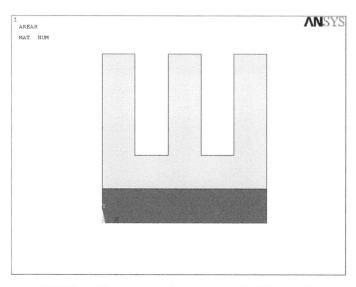

ANSYS graphics shows the fin and chip with different color

Main Menu > Preprocessor > Meshing > Mesh Tool

A select Smart Size

B set the level to 1

C Mesh

Click on Pick All

Close

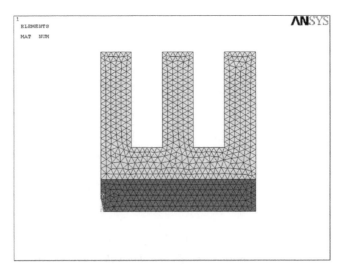

ANSYS graphics shows the mesh

Boundary conditions are applied for the solution. A convective boundary condition is applied at the fin surface, and at the vertical sides of the chip. The bottom surface of the chip is well insulated, and the zero heat flux simulates the insulation boundary condition. If no boundary condition is specified at the external surfaces, the ANSYS will consider it, by default, as an insulated boundary condition. No boundary condition is applied at the common line between the chip and the fin. Finally, a volumetric heat generation is applied at the chip.

Main Menu > Solution > Define Load > Apply > Thermal > Convection > On Lines

Click on fin surfaces and the vertical surface of the chip where convection boundary condition is applied.

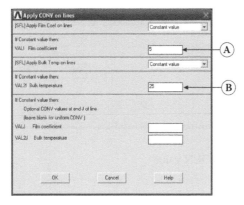

A type 5 in Film Coefficient

B type 25 in Bulk temperature

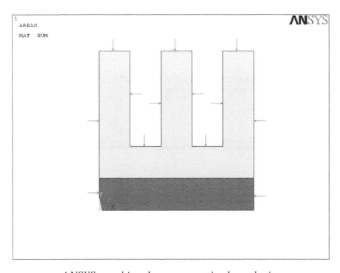

ANSYS graphics shows convective boundaries

The heat generation must be per unit volume. The applied heat generation is divided by area of the chip because the problem is two dimensional. The chip volumetric heat generation is calculated as follows:

$$\dot{Q} = \frac{15}{0.05 \times 0.01} = 30,000 \text{ W/m}^2$$

Main Menu > Solution > Define Load > Apply > Thermal > Heat Generat > On Area

Click on the chip area, where heat generation is applied.

OK

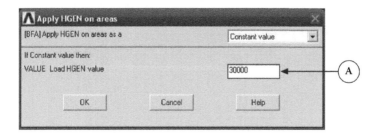

A type 30000 in Load HGEN value

OK

The preprocessor and solution tasks are now completed. Now, the problem can be solved.

Main Menu > Solution > Solve > Current LS

OK

| Close |

The solution is completed successfully, and no error message is posted. In the post-processor, the contours of the temperature should be carefully inspected to ensure that the boundary conditions are applied correctly and there are no mistakes in the preprocessor, such as wrong input of the dimensions or material properties.

Main Menu > General Postproc > Plot Results > Contour Plot > Nodal Solu

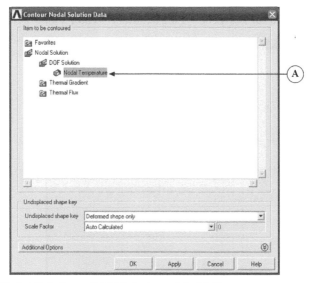

A click on Nodal Solution > DOF Solution > Nodal Temperature

| OK |

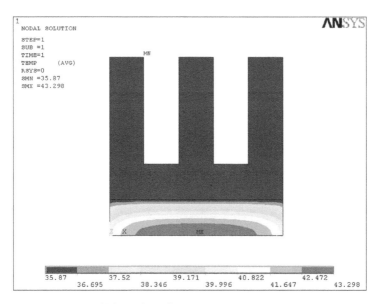

ANSYS graphics shows temperature contours

The temperature contours indicate that the maximum temperature is located at the bottom surface of the chip, which is equal to 43.298°C. The default number of contours is 9, and this number can be increased to a higher value for better data analysis. First, the graphics device must be changed to win32C, and the number of contours can be increased up to 114 contours. Notice that increasing the number of contours does not mean that the accuracy of the result is improved. The vector plot showing the heat flow from the chip to the fins is presented. The red arrow is for high value of the heat flux.

Main Menu > General Postproc > Plot Results > Vector Plot > Predefined

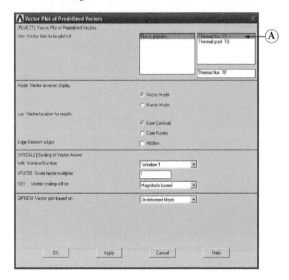

A select Thermal flux TF

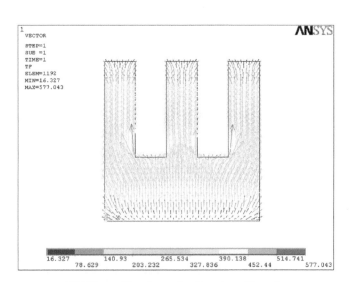

ANSYS graphics shows the direction of heat flux

The average temperature at the base of the chip is calculated using the path operation. To create a path, there are two options: Arbitrary and Circular paths. The Arbitrary path can be made from the line segments by clicking on the ANSYS graphics, and the grids should be enabled, and the Circular path has a circular path; a center point and radius are specified. For this example, the Arbitrary path is utilized.

Main Menu > General Postproc > Path Operation > Define Path > On Working Plane

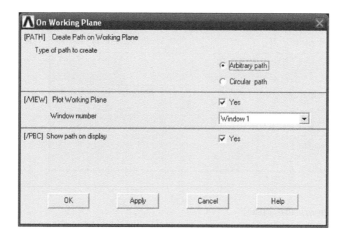

Using the mouse, click on the ANSYS graphics window at the right and left bottom corners of the chip.

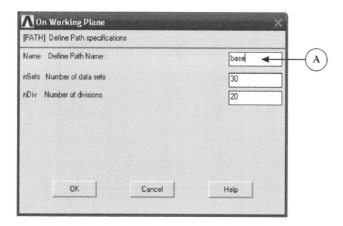

A name the path as base

The name of the path is optional. The number of data set is the maximum number of field variables. The number of divisions is 20 by default, and increasing this number to 50 will produced a smother plot. Next, the field variable is assigned to the path for plotting. This can be accomplished by using the Map onto Path in the path operation. Only one variable can be selected. For this example, the temperature is selected.

Main Menu > General Postproc > Path Operation > Map onto Path

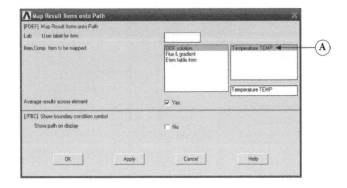

A select Temperature TEMP

Now, the stored variable is ready to be plotted. In the Plot path item, there are two options. The stored data can be either plotted on a graph or listed. The list results can be exported to other graphical software such as EXCEL.

Main Menu > General Postproc > Path Operation > Plot path Item > On Graph

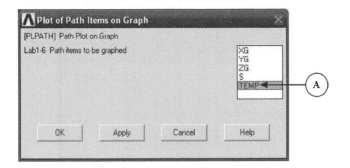

A select TEMP

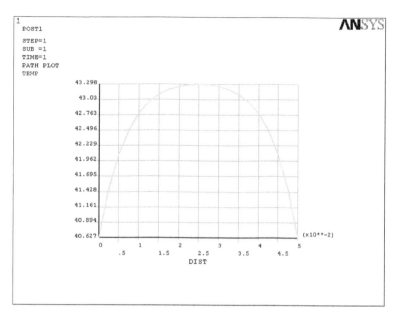

ANSYS graphics shows channel temperature at the base

The temperature at the base is perfectly parabolic due to the symmetry in the problem. The above graph indicates that the maximum temperature at the bottom surface of the chip is 43.298°C. The average temperature at the base can be determined using the integration in the path operation. The value of the integration must be divided by the path length to get the average value of the variable, and the path length is 0.05 m. Hence, number 20 is entered in the FACT that will be multiplied by the integration result. Selecting S in the Lab2 means that the integration is performed along the path.

Main Menu > General Postproc > Path Operation > Integrate

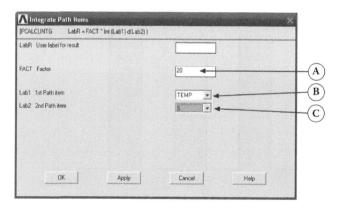

A type 20

B select TEMP

C select S

OK

```
ANSYS 10.0 Output Window                              _ □ ×
 DIRECTION        MAX                MIN
     X        0.50000E-01        0.0000
     Y           0.0000          0.0000
     Z           0.0000          0.0000
 TOTAL PATH LENGTH =    0.50000E-01

 DEFINE PATH VARIABLE TEMP      AS THE DEGREE OF FREEDOM ITEM=TEMP COM
   NUMBER OF PATH VARIABLES DEFINED IS    5
 SUMMARY OF VARIABLE TEMP      MAX =     43.298    MIN =     40.627

 PATH BOUNDARY CONDITION DISPLAY KEY =  0

 DISPLAY ALONG PATH DEFINED BY LPATH COMMAND.  DSYS= 0
 TURN OFF WORKING PLANE DISPLAY

 DEFINE PATH VARIABLE          AS THE INTEGRATION OF
   PATH VARIABLE TEMP      WITH RESPECT TO PATH VARIABLE S
   FINAL SUMMATION =    42.666
   NUMBER OF PATH VARIABLES DEFINED IS    6
```

ANSYS Output window shows the value of the average temperature at the base which is 42.666°C.

4.4 UNSTEADY THERMAL ANALYSES OF FIN

For the fin shown in Figure 4.4, solve the problem as an unsteady state and determine the temperature distribution in the fin at 100 s if the initial temperature of the fin is 25°C. Plot the temperature history at point A. Also, create a temperature animation file for the heating process. The total duration is 200 s and the time step is 2 s. The fin is made of nickel–steel (10%) with the following properties: $\rho = 7945\,\text{kg/m}^2$, $k = 26\,\text{W/m-°C}$, and $C_p = 4600\,\text{J/kg-°C}$. The bottom surface of the fin is maintained at 100°C, and the surface is subjected to free convection with $h = 5\,\text{W/m-°C}$ and 25°C.

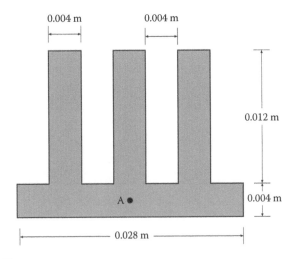

FIGURE 4.4   Fin.

Double click on the ANSYS icon

Main Menu > Preferences

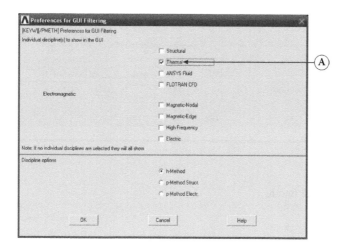

A select the Thermal

OK

Main Menu > Preprocessor > Element type > Add/Edit/Delete

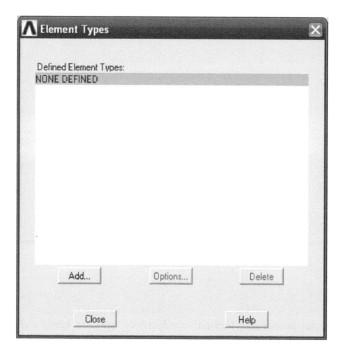

Add...

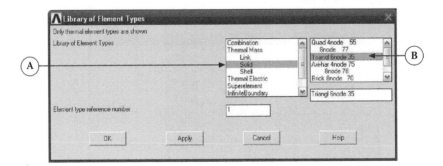

A select Solid

B select Triangle 6node

OK

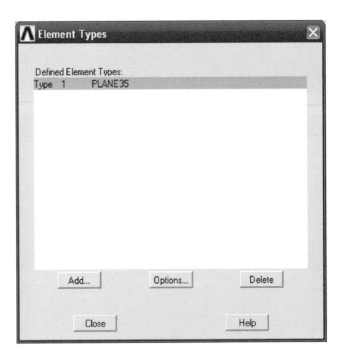

Close

For the material properties, thermal conductivity, specific heat, and density are required to solve the problem because the problem is unsteady.

Main Menu > Preprocessor > Material Props > Material Models

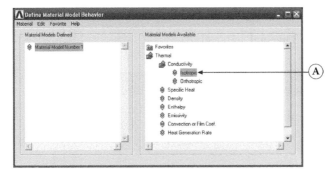

A double click on Thermal > Conductivity > Isotropic

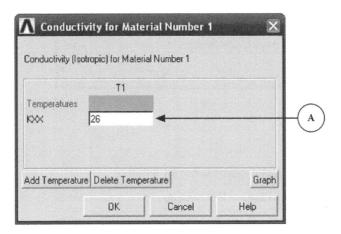

A type 26 in KXX

OK

Double click on Thermal > Specific Heat

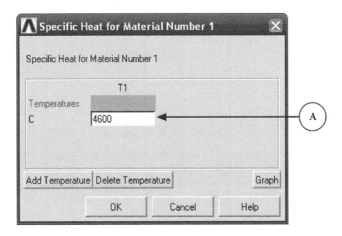

A type 4600 in C

OK

Double click on Thermal > Density

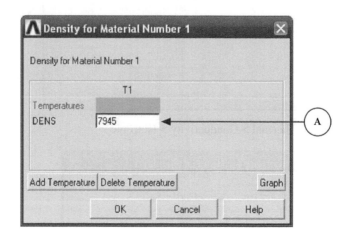

A type 7945 in DENS

OK

Close the define material model behavior window

Utility Menu > WorkPlane > WP Setting

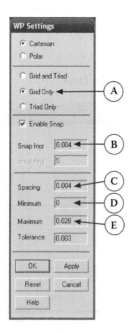

A select Grid Only

B type 0.004 in Snap Incr

C type 0.004 in Spacing

D type 0 in Minimum

E type 0.028 in Maximum

ANSYS Utility Menu > WorkPlane > Display Working Plane

ANSYS Utility Menu > PlotCtrls > Pan, Zoom, Rotate

Click on zoom in and zoom out, and the cursor buttons until the ANSYS graphics shows all grids.

ANSYS Main Menu > Preprocessor > Modeling > Create > Key points > On Working Plane

Using the mouse, click at the location Key points as shown in the figure below.

ANSYS graphics shows the Key points

ANSYS Main Menu > Preprocessor > Modeling > Create > Lines > Lines > Straight Line

Click on two Key points to form one straight line, and redo for all lines, as shown below.

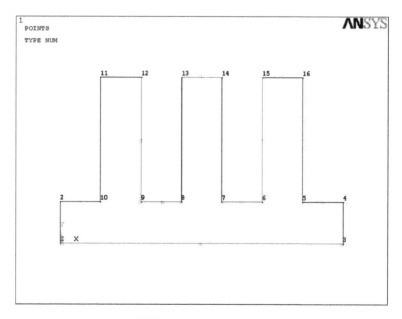

ANSYS graphics shows the lines

ANSYS Main Menu > Preprocessor > Modeling > Create > Areas > Arbitrary > By Lines

Click on all lines to create an area for the fin.

OK

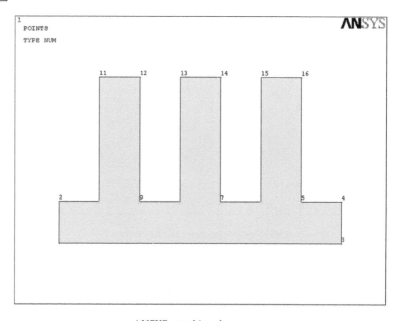

ANSYS graphics shows area

Main Menu > Preprocessor > Meshing > Mesh Tool

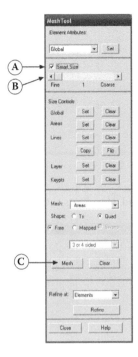

A select Smart Size

B set the level to 1

C Mesh

Click on **Pick All**

Close

The type of the analysis will be changed to transient, and a full solution method is selected. This solution method imposes the boundary conditions at time zero. DB/ Result file is used to control the number of output results to be stored for the postprocessor. Data will be stored at each time step during the solution with every subset option. The total duration for the simulation and time step is specified in the Time/Frequency command. In the time–time step windows, make sure that "stepped" is selected.

Main Menu > Solution > Analysis Type > New Analysis

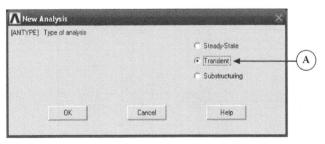

A select Transient

OK

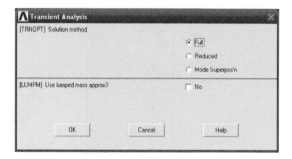

OK

Main Menu > Solution > Load Step Opts > Output Ctrls > DB/Results File

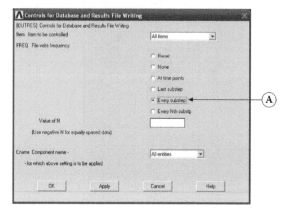

A select Every subset

OK

Main Menu > Solution > Load Step Opts > Time/Frequenc > Time –Time Step

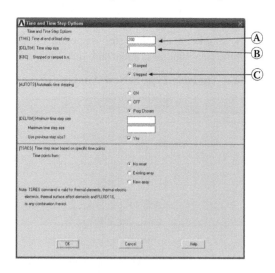

A type 200 in the Time at end of load step

B type 1 in Time step size

C select Stepped

 OK

The 1 s is selected for the time step size to have a total of 200 time steps for this transient problem. If less time step size is used, such as 0.5 s, the results will be more accurate, but the computational time will be doubled. Boundary conditions are applied in the solution task. Temperature and convection boundary conditions are applied at the external surface. The upper surfaces are subjected to convection. The bottom surface is maintained at a constant temperature.

Main Menu > Solution > Define Load > Apply > Thermal > Convection > On Lines

Click on fin surfaces where a convective boundary condition is applied.

 OK

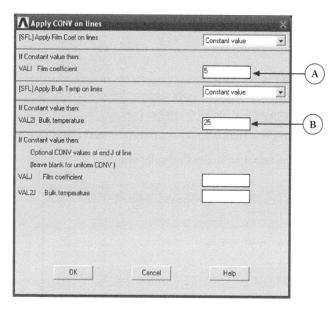

A type 5 in Film Coefficient

B type 25 in Bulk temperature

 OK

Main Menu > Solution > Define Load > Apply > Thermal > Temperature > On Lines

Click on fin bottom surface where a temperature boundary condition is applied.

 OK

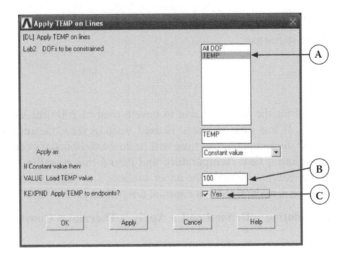

A select TEMP

B type 100 in Load TEMP value

C select Yes

OK

When Yes is selected in KEXPND, the temperature is applied at the selected line, as well as the key points of that line. Imposing the boundary conditions is now completed. The initial condition is specified as follows:

Main Menu > Solution > Define Load > Apply > Initial Condition > Define

Click on **Pick All** to select all nodes in the domain

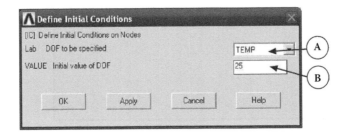

A select TEMP in the DOF to be specified

B type 25 in Initial value of DOF

OK

The solution task is now completed, and the model is now ready to be solved. During the solution task, the ANSYS output windows will show the progress of the solution. Carefully monitor the run.

Main Menu > Solution > Solve > Current LS

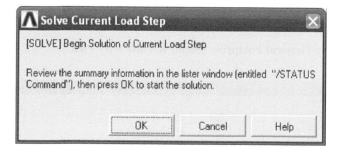

OK

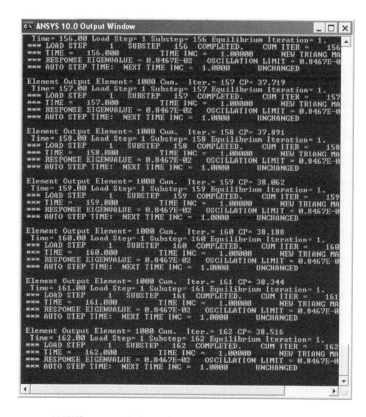

ANSYS output windows shows transient solution progress

Close

The temperature contours will be presented at time steps of 100 s. First, the time step is loaded, then the temperature contours are plotted.

Main Menu > General Postproc > Read Results > By Pick

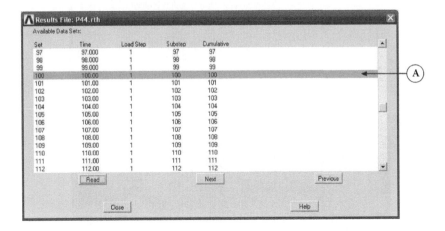

A select Time 100.00

Read and **Close**

Main Menu > General Postproc > Plot Results > Contour Plot > Nodal Solu

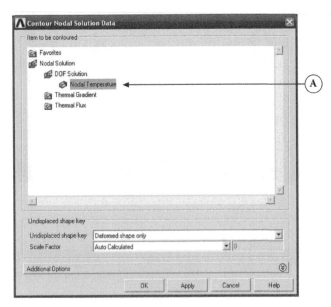

A click on Nodal Solution > DOF Solution > Nodal Temperature

OK

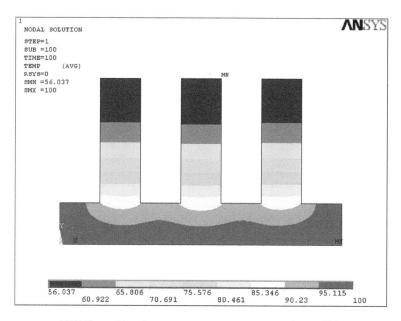

ANSYS graphics shows temperature contours at time = 100 s

Determining the temperature history at a specific location in the domain is important. Here, the temperature history at the center of the fin's base is presented in graphical form. The steps are shown below:

Main Menu > TimeHist Postpro

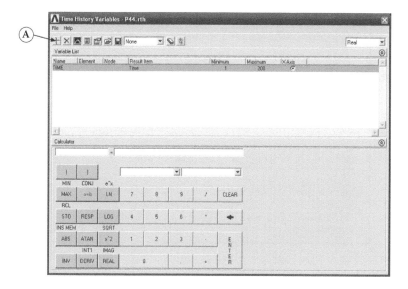

A Click on the green X

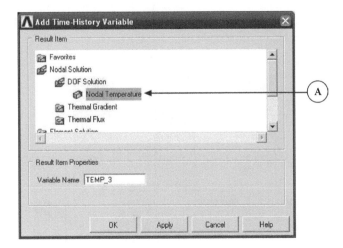

A click on Nodal solution > DOF Solution > Nodal Temperature

$\boxed{\text{OK}}$

Using the mouse, click at the point A. ANSYS will not accept double click. The results can be presented either in graphical or list form. The graphical presentation is selected for this exercise. Carefully watch the temperature curve. It starts from the specified initial condition to reach the steady-state condition.

$\boxed{\text{OK}}$

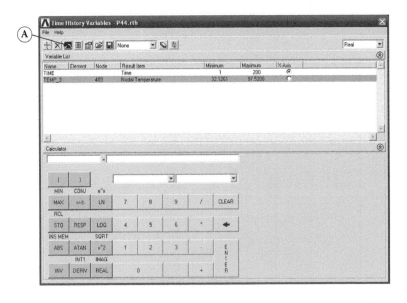

A click on the graph

$\boxed{\text{OK}}$

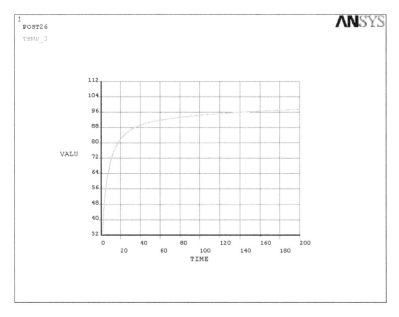

ANSYS graphics shows temperature history of the selected location

Animation of the temperature contours from time = 0 to 200 s can be easily accomplished using animate command in the PlotCtrol options. The number of the frames in the animate over time is the number of pictures in the avi file, while the animation time delay is the display period between the two pictures.

Main Menu > General Postproc

Utility Menu > PlotCtrls > Animate > Over time …

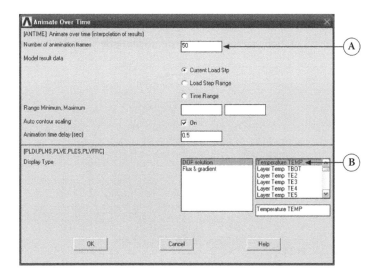

A type 50 in Number of animation frames

B select temperature

The ANSYS will show an animation of the heating process of the fin. The animation file will be stored in the working directory, and its format is avi.

4.5   PHASE CHANGE HEAT TRANSFER

A thermal storage device with a phase change material (PCM) is commonly used in thermal systems to store heat. The stored heat in the PCM can then be used during high-energy demand periods to conserve energy. The PCM can be paraffin wax, salt hydrate, or even water. Consider the PCM system shown in Figure 4.5 that consists of a metallic pipe containing paraffin wax (n-Eicosane). The outer diameter of the pipe is subjected of constant temperature boundary condition of 100°C. The initial temperature for the entire system is 25°C. The n-Eicosane wax is initially in the solid state. Find the temperature history at the center of the wax cylinder during the interval of $0 < \text{time} < 2500\,\text{s}$, and estimate the time required to completely melt the wax. Given $R_i = 0.02\,\text{m}$ and $R_o = 0.022\,\text{m}$.

Pipe	**n-Eicosane wax**
$\rho = 1200\,\text{kg/m}^3$	$\rho = 1000\,\text{kg/m}^3$
$k = 50\,\text{W/kg-°C}$	$k = 0.5\,\text{W/kg-°C}$
$C_p = 900\,\text{J/kg-°C}$	$C_p = 1050\,\text{J/kg-°C}$
	$\lambda = 210\,\text{kJ/kg}$
	$T_m = 36\text{°C}$

where λ and T_m are the latent heat of fusion and the melting temperature of the n-Eicosane wax, respectively. For simplicity, thermal expansion of the PCM and pipe is not considered in the simulations. The thermal expansion analysis will highly complicate the computations and it has an insignificant effect on the heat transfer in

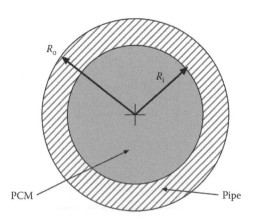

FIGURE 4.5   PCM and pipe system.

the system. The problem is solved as a heat conduction problem, and therefore, the effect of the natural convection of liquid PCM is not considered.

Numerical techniques, such as finite element method, are commonly utilized to solve phase change heat transfer problems. For modeling phase change problems, there are two main approaches: moving-mesh and fixed-mesh methods. The mesh of the moving-mesh method is allowed to change to track the solid/liquid interface. This method is rarely used in practice because it will highly complicate the computations. The fixed-mesh method is commonly used for modeling phase change. In this method, the geometry of the mesh is independent of time, and the liquid/solid interface is tracked by the definition of the specific heat in the governing equations. The specific heat is the rate of change of enthalpy with respect to temperature:

$$C_p = \frac{\lambda}{\delta T} \tag{4.33}$$

The apparent specific heat is the sum of the sensible and latent heats, which is defined as

$$\bar{C}_p = C_p + \frac{\lambda}{\delta T} \tag{4.34}$$

The term $\lambda/\delta T$ is equal to zero at the solid and liquid phases. The δT is equal to 1°C. The following is the definition of the apparent specific heat used to model the phase change:

$$\bar{C}_p = \begin{cases} C_p & T \leq T_m \\ C_p + \dfrac{\lambda}{\delta T} & T_m < T < T_m + \delta T \\ C_p & T \leq T_m + \delta T \end{cases} \tag{4.35}$$

To approximate the latent heat effect, the relationship between the temperature and the specific heat for the PCM in Figure 4.6 is used in the ANSYS.

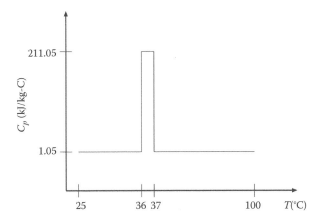

FIGURE 4.6 Specific heat as a function of temperature for the PCM.

Double click on the ANSYS icon

Main Menu > Preferences

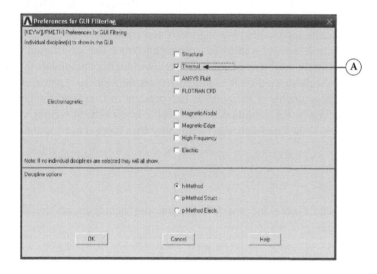

 A select the Thermal

OK

Main Menu > Preprocessor > Element type > Add/Edit/Delete

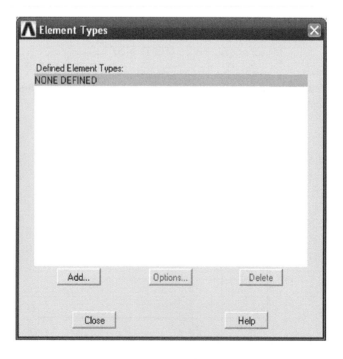

Add...

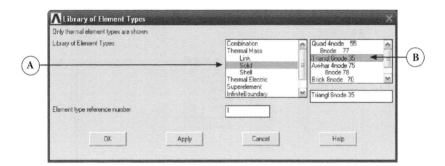

A select Solid

B select Triangle 6node

OK

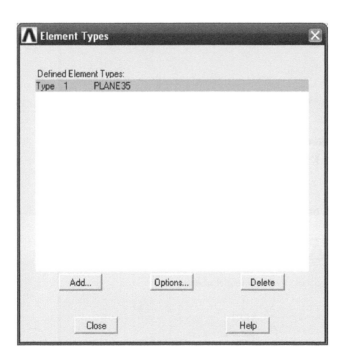

Close

In this example, the specific heat of the PCM is function of temperature. The value of the property and the corresponding temperature value can also be plotted. Material number 1 is selected for the pipe, while material number 2 is for the PCM.

Main Menu > Preprocessor > Material Props > Material Models

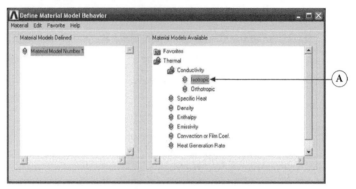

A double click on Thermal > Conductivity > Isotropic

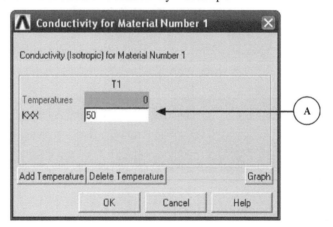

A type 50 in KXX

OK

Double click on Thermal > Specific Heat

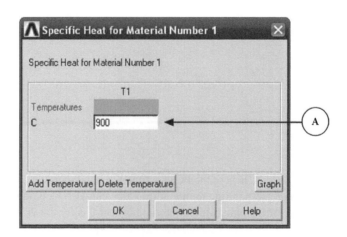

A type 900 in C

OK

Double click on Thermal > Density

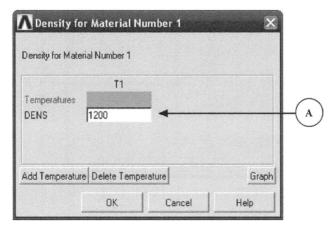

A type 1200 in DENS

OK

In the Define Material Models Behavior: Material > New Model

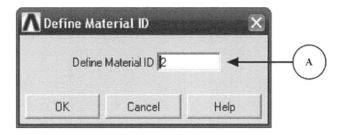

A type 2 in Define Material ID

OK

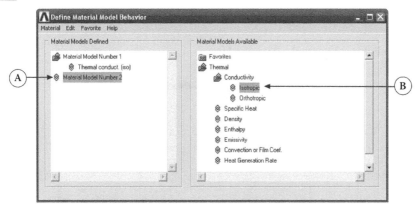

A Select Model Number 2

B double click on Thermal > Conductivity > Isotropic

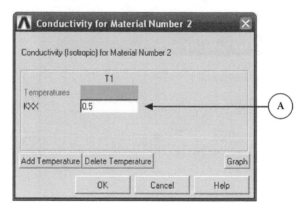

A type 0.5 in Temperature

$\boxed{\text{OK}}$

Double click on Thermal > Density

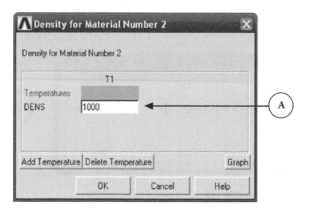

A type 1000 in DENS

$\boxed{\text{OK}}$

Double click on Thermal > Specific Heat

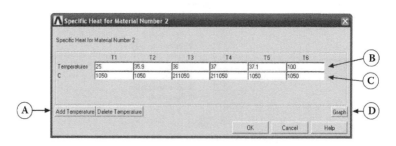

A Click on the Add Temperature for five times

B type 25, 35.9, 36, 37, 37.1, and 100 in Temperature

C type 1050, 1050, 211050, 211050, 1050, and 1050 in C

D Click on Graph

OK

Close the material model behavior window

By clicking on the Graph, a plot of specific heat versus temperature will be shown in the ANSYS graphics, as shown below. Carefully examine the graph to avoid any wrong input.

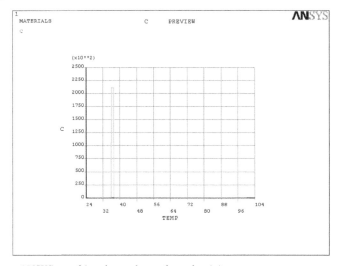

ANSYS graphics shows thermal conductivity versus temperature

The modeling task is started here. First, two solid circles with radius 0.022 and 0.02 are created. Then, the circles are overlapped.

Main Menu > Preprocessor > Modeling > Create > Areas > Circle > Solid Circle

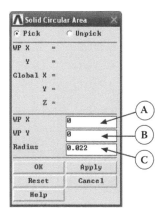

A type 0 in WP X

B type 0 in WP Y

C type 0.022 in Radius

OK

Main Menu > Preprocessor > Modeling > Create > Areas > Circle > Solid Circle

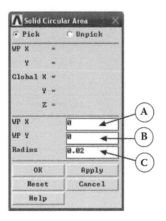

A type 0 in WP X

B type 0 in WP Y

C type 0.02 in Radius

OK

Main Menu > Preprocessor > Modeling > Operate > Booleans > Overlap > Areas

Pick All

Main Menu > Preprocessor > Meshing > Mesh Tool

A select Area

B click on Set

Using mouse, select the inner circle.

OK

Area attribute windows will show up. Select number 2, corresponding to material number for the wax.

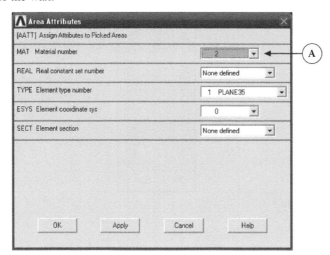

A select number 2 in Material number

OK

Utility Menu > PlotCtrls > Numbering...

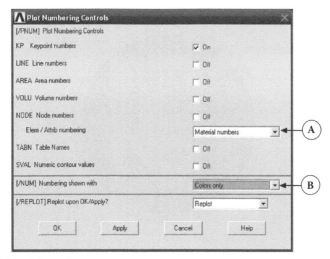

A select Material number

B select Colors only

OK

Main Menu > Preprocessor > Meshing > Mesh Tool

A select Smart Size

B set the level to 1

C Mesh

Click on $\boxed{\textbf{Pick All}}$

$\boxed{\textbf{Close}}$

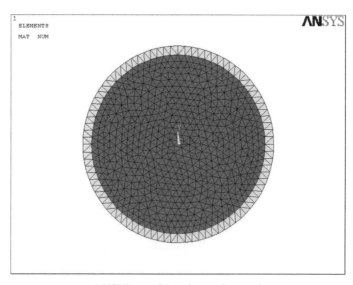

ANSYS graphics shows the mesh

Main Menu > Solution > Analysis Type > New Analysis

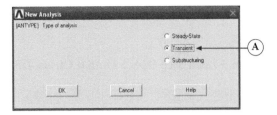

A select Transient

OK

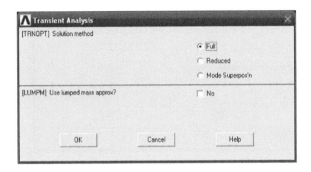

OK

The relationship between the specific heat and temperature for the PCM is nonlinear. If the full scheme is used, the solution will not be converged. The linear scheme should not also be used because it is not capable to simulate the given relationship between the specific heat and temperature. The quasi scheme should be used. The equation solver should be changed to Jacobi Conj Grad. To store the temperature field data at all time steps, the every substep should be selected.

Main Menu > Solution > Analysis Type > Analysis Options

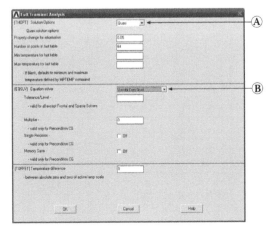

A select Quasi in Solution Options

B select Jacobi Conj Grad In Equation solver

Main Menu > Solution > Load Step Opts > Output Ctrls > DB/ Results File

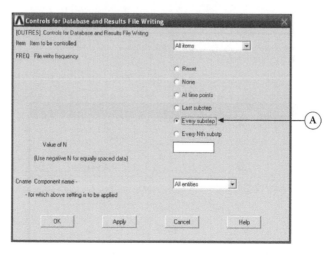

A select Every subset

Main Menu > Solution > Load Step Opts > Time/Frequency> Time-time step

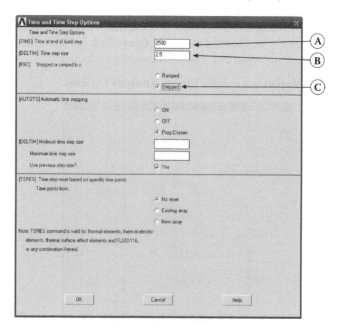

A type 2500 in the Time at end of load step

B type 2.5 in Time step size

C select Stepped

 OK

The 2.5 s is selected for the time step size to have 1000 time steps for this transient problem. If less time step size is used, the results will be more accurate, but the computational time will be more.

Main Menu > Solution > Define Load > Apply > Thermal > Temperature > On Lines

Click on all lines of the outer surface of the pipe where temperature boundary condition is applied.

OK

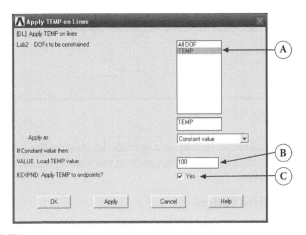

A select TEMP

B type 100 in Load TEMP value

C click on Yes

OK

Main Menu > Solution > Define Load > Apply > Initial Condition > Define

Pick All

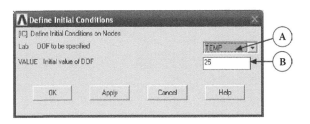

A select TEMP

B type 25 in Initial value of DOF

OK

Main Menu > Solution > Solve > Current LS

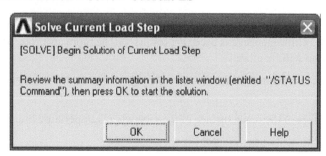

OK

Close

Determining the temperature history at a specific location in the domain is important. Here, the temperature history at the center of the wax is presented in a graphical form. The steps are shown below:

Main Menu > TimeHist Postpro

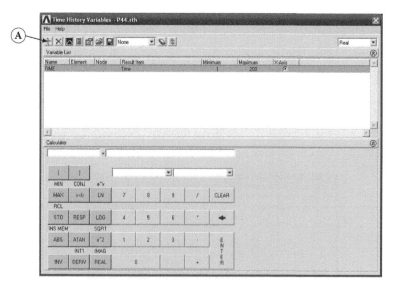

A Click on the green X

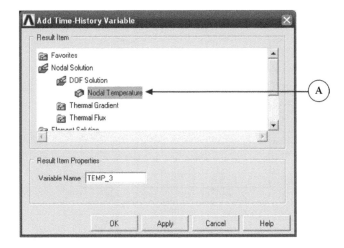

A click on Nodal Solution > DOF Solution > Nodal Temperature

OK

Using the mouse, click at the center of the wax cylinder. ANSYS will not accept double click, an error message will be shown. Carefully watch the temperature curve. It starts from the specified initial condition to reach the steady-state condition.

OK

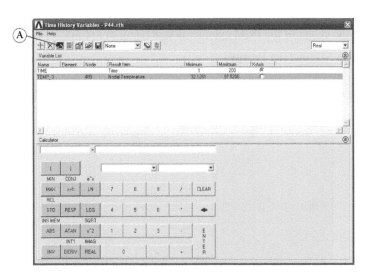

A Click on the graph

OK

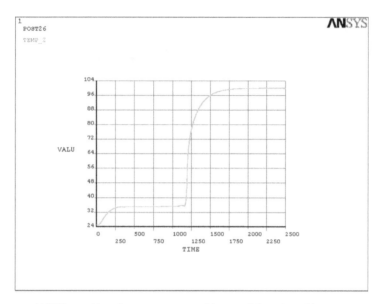

ANSYS graphics shows temperature history of the selected location

The above figure reveals important phenomena. The temperature starts to increase from its initial temperature of 25°C, showing correct specified initial temperature. Then, the temperature increases until it reaches 36°C, which is the PCM melting temperature. At this point, the specific heat of the PCM increases to more than 200 times which makes the temperature almost constant and high amount of energy is absorbed by the PCM. Once the PCM is completely melted at time = 1150 s, the PCM temperature is increased drastically until it reaches the steady-state temperature. In order to estimate the time required to melt the PCM, a list of the temperature history at the center of the PCM cylinder is generated, as follows:

Main Menu > TimeHist Postpro

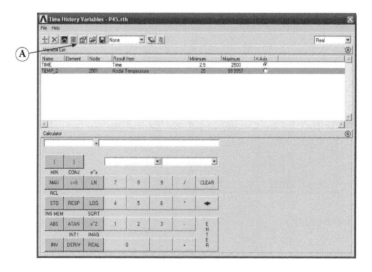

A Click on the list button, the button next to the graph button

OK

PRVAR Command		
File		
1147.5	35.6029	
1150.0	35.3674	

***** ANSYS POST26 VARIABLE LISTING *****

TIME	2501 TEMP
	TEMP_2
1152.5	35.2352
1155.0	35.3098
1157.5	35.4278
1160.0	35.5760
1162.5	35.7581
1165.0	35.9252
1167.5	36.1360
1170.0	36.6401
1172.5	37.1856
1175.0	37.9028
1177.5	38.7829
1180.0	40.0162
1182.5	41.8372
1185.0	44.4468
1187.5	45.3785
1190.0	47.0399
1192.5	53.3549

The PCM is completely melted when the temperature at its center reached 37°C. From the above table, the PCM is completely melted at 1172.05 s.

PROBLEMS

4.1 Study the thermal performance of a straight fin used for an electronic chip by determining the maximum operating temperature of the chip, as shown in Figure 4.7. The fin is made of pure aluminum. Forced convection boundary condition is imposed at the fin surface with $h = 25$ W/m²-°C, $T_0 = 23.5$°C, and the convection is also applied at the chip vertical surfaces, but its base is insulated. A power of 35 W is generated in the chip. Let $k_{chip} = 0.15$ W/m-°C and $k_{Al} = 237$ W/m-°C. Solve the problem as steady state. Use triangle 6node element with smart size of 1.

4.2 Study the thermal performance of the straight fin used for electronic chip, as shown in Figure 4.8. The fin is made of pure aluminum. Forced convection boundary condition is imposed at the fin surface with $h = 15$ W/m²-°C, $T_0 = 22$°C, and the convection is also applied at the chip surfaces, but its base is insulated. A power of 20 W is generated in the chip. Let $k_{chip} = 0.15$ W/m-°C and $k_{Al} = 230$ W/m-°C. Solve the problem as steady state. Use triangle 6node element with smart size of 1. Determine

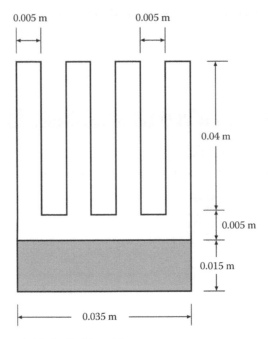

FIGURE 4.7   Fin and chip for Problem 4.1.

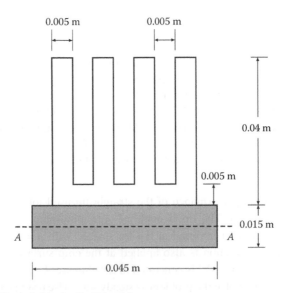

FIGURE 4.8   Fin and chip for Problem 4.2.

a. Maximum operating temperature
b. Temperature distribution along path A–A
c. Average temperature along path A–A

4.3 Study the thermal performance of the straight fin used for electronic chip, as shown in Figure 4.9. The fin is made of pure aluminum. Forced convection boundary condition is imposed at the fin surface with $h = 15\,W/m^2\text{-}°C$, $T_0 = 22°C$, and the convection is also applied at the chip surfaces, but its base is insulated. A power of 15 W is generated in the heat source only. Let $k_{chip} = 0.2\,W/m\text{-}°C$ and $k_{Al} = 230\,W/m\text{-}°C$. Solve the problem using the ANSYS as steady state. Use triangle 6node element with smart size of 1. Determine

1. Maximum operating temperature
2. Temperature at the center of the heat source
3. Temperature distribution along path A–A
4. Average temperature along path A–A

4.4 For the electronic board with mounted macro-processor and thermally inactive memory shown in Figure 4.10, determine the maximum operating temperature of the processor. In addition, present the temperature contours and heat flux distribution. The applied heat generation at the processor is 12 W. Convection boundary condition is imposed ($h = 7.5\,W/m^2\text{-}°C$ and $T_\infty = 25°C$) at the top surface, while it is insulated at the bottom surface and at the sides of the IC board. Use triangle 6node element with smart size of 1.

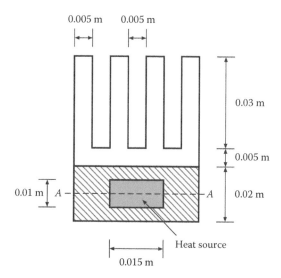

FIGURE 4.9   Fin and chip for Problem 4.3.

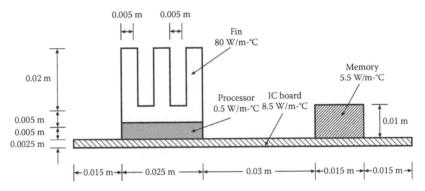

FIGURE 4.10 Electronic system for Problem 4.4.

4.5 Solve Problem 4.3 if the thermal conductivity of the chip is function of temperature, and the relationship is given in Figure 4.11.

4.6 For the fin and chip shown in Problem 4.1, solve the problem as an unsteady and determine the maximum temperature of the device at 1000 and 2000 s, and determine the temperature–time history at the center of the chip. Also, create temperature animation file for the heating process. The initial temperature of the fin and chip is 23.5°C. The total duration is 2000 s and the time step 20 s. The fin is made of nickel–steel (10%) with the following properties: $\rho = 7945\,kg/m^2$, $k = 26\,W/m\text{-}°C$, and $C_p = 4600\,J/kg\text{-}°C$. The chip is made of epoxy with the following properties: $\rho = 1200\,kg/m^2$, $k = 0.25\,W/m\text{-}°C$, and $C_p = 1000\,J/kg\text{-}°C$.

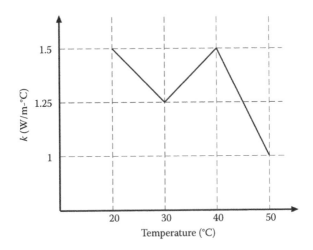

FIGURE 4.11 Thermal conductivity of the chip for Problem 4.5.

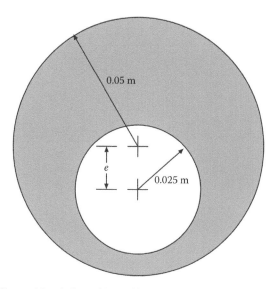

FIGURE 4.12 Thermal insulation with small eccentricity.

4.7 The radial thermal insulation, as shown in Figure 4.12, has a small eccentricity. Determine the effect of the eccentricity on the overall thermal resistance of the insulation material. The inner and outer surfaces are maintained at temperatures 75°C and 25°C, respectively. The thermal conductivity of the insulation material is 0.25 W/m-K. Create an $X–Y$ plot showing the eccentricity in the x-axis and the heat flux at the outer surface in the y-axis. Consider eccentricity between $e = 0$ and $e = 0.025$.

5 Fluid Mechanics

5.1 GOVERNING EQUATIONS FOR FLUID MECHANICS

The mass conservation equation in a differential form can be obtained by applying the mass conservation principle on a differential control volume, as shown in Figure 5.1. Considering the elemental Cartesian fixed control volume, the net mass flow rate in the x-, y-, and z-directions can be expressed as

$$x\text{-direction}: \frac{\partial}{\partial x}(\rho u)\, dx\, dy\, dz \tag{5.1}$$

$$y\text{-direction}: \frac{\partial}{\partial y}(\rho v)\, dx\, dy\, dz \tag{5.2}$$

$$z\text{-direction}: \frac{\partial}{\partial z}(\rho w)\, dx\, dy\, dz \tag{5.3}$$

The rate of change of mass inside the control volume can be obtained from Reynolds transport theory as follows:

$$\int_{CV} \frac{\partial \rho}{\partial t}\, dV = \frac{\partial \rho}{\partial t}\, dx\, dy\, dz \tag{5.4}$$

The net mass flux into the control volume should be equal to the rate of change of mass inside the control volume. The mass conservation in differential form can be expressed as

$$\frac{\partial \rho}{\partial t} + \frac{\partial}{\partial x}(\rho u) + \frac{\partial}{\partial y}(\rho v) + \frac{\partial}{\partial z}(\rho w) = 0 \tag{5.5}$$

Newton's second law on a differential control volume, as shown in Figure 5.1, can be used to obtain the conservation of momentum equation. The net forces on the control volume should be balanced with the acceleration of the control volume times its mass as follows:

$$\vec{a}\, dm = \sum d\vec{F} \tag{5.6}$$

245

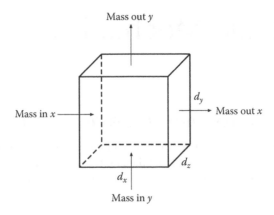

FIGURE 5.1   Fixed control volume.

The acceleration vector field, $\vec{a}$, can be obtained from the total time derivative of the velocity vector:

$$\vec{a} = \frac{dV}{dt} = \frac{du}{dt}i + \frac{dv}{dt}j + \frac{dw}{dt}k \tag{5.7}$$

Each component of the velocity fields is a function of space and time variables. Using the chain rule, the time derivative can be obtained:

$$\frac{du(x,y,z,t)}{dt} = \frac{\partial u}{\partial t} + \frac{\partial u}{\partial x}\frac{dx}{dt} + \frac{\partial u}{\partial y}\frac{dy}{dt} + \frac{\partial u}{\partial z}\frac{dz}{dt} \tag{5.8}$$

where
   $u = dx/dt$ is the local velocity component in the x-direction
   $v = dy/dt$ is the local velocity component in the y-direction
   $w = dw/dt$ is the local velocity component in the z-direction

The total derivative of u is the acceleration in the x-direction:

$$a_x = \frac{du(x,y,z,t)}{dt} = \frac{\partial u}{\partial t} + u\frac{du}{dx} + v\frac{du}{dy} + w\frac{du}{dz} \tag{5.9}$$

The acceleration in the y- and z-directions can be expressed, respectively, as

$$a_y = \frac{dv(x,y,z,t)}{dt} = \frac{\partial v}{\partial t} + u\frac{dv}{dx} + v\frac{dv}{dy} + w\frac{dv}{dz} \tag{5.10}$$

$$a_z = \frac{dw(x,y,z,t)}{dt} = \frac{\partial w}{\partial t} + u\frac{dw}{dx} + v\frac{dw}{dy} + w\frac{dw}{dz} \tag{5.11}$$

The mass of the control volume must be equal to volume times the density:

$$dm = \rho\, dx\, dy\, dz \tag{5.12}$$

The forces on the control volume are divided into two types: body and surface forces. The body force is due to gravity:

$$dF_b = \rho \vec{g}\, dx\, dy\, dz \tag{5.13}$$

The surface force is due to the surface stresses, including normal and parallel stresses. The surface stresses in the x-, y-, and z-directions are as follows:

$$dF_{sx} = \left(\frac{\partial \sigma_{xx}}{\partial x} + \frac{\partial \tau_{yx}}{\partial y} + \frac{\partial \tau_{zx}}{\partial z} \right) dx\, dy\, dz \tag{5.14}$$

$$dF_{sy} = \left(\frac{\partial \sigma_{yy}}{\partial y} + \frac{\partial \tau_{xy}}{\partial y} + \frac{\partial \tau_{zy}}{\partial z} \right) dx\, dy\, dz \tag{5.15}$$

$$dF_{sz} = \left(\frac{\partial \sigma_{zz}}{\partial x} + \frac{\partial \tau_{xz}}{\partial y} + \frac{\partial \tau_{yz}}{\partial z} \right) dx\, dy\, dz \tag{5.16}$$

The equations of motion in the x-, y-, and z-directions can be expressed as

$$x\text{-direction}: \rho\left(\frac{\partial u}{\partial t} + u\frac{\partial u}{\partial x} + v\frac{\partial u}{\partial y} + w\frac{\partial u}{\partial z} \right) = \left(\rho g_x + \frac{\partial \sigma_{xx}}{\partial x} + \frac{\partial \tau_{yx}}{\partial y} + \frac{\partial \tau_{zx}}{\partial z} \right) \tag{5.17}$$

$$y\text{-direction}: \rho\left(\frac{\partial v}{\partial t} + u\frac{\partial v}{\partial x} + v\frac{\partial v}{\partial y} + w\frac{\partial v}{\partial z} \right) = \left(\rho g_y + \frac{\partial \sigma_{yy}}{\partial x} + \frac{\partial \tau_{xy}}{\partial y} + \frac{\partial \tau_{zy}}{\partial z} \right) \tag{5.18}$$

$$z\text{-direction}: \rho\left(\frac{\partial w}{\partial t} + u\frac{\partial w}{\partial x} + v\frac{\partial w}{\partial y} + w\frac{\partial w}{\partial z} \right) = \left(\rho g_z + \frac{\partial \sigma_{zz}}{\partial x} + \frac{\partial \tau_{xz}}{\partial y} + \frac{\partial \tau_{yz}}{\partial z} \right) \tag{5.19}$$

For a Newtonian fluid, the stress components can be obtained from the theory of elasticity, and they are

$$\sigma_{xx} = -P + 2\mu \frac{\partial u}{\partial x} \qquad (5.20)$$

$$\sigma_{yy} = -P + 2\mu \frac{\partial v}{\partial v} \qquad (5.21)$$

$$\sigma_{zz} = -P + 2\mu \frac{\partial w}{\partial z} \qquad (5.22)$$

$$\tau_{xy} = \tau_{yx} = \mu \left(\frac{\partial u}{\partial y} + \frac{\partial v}{\partial x} \right) \qquad (5.23)$$

$$\tau_{yz} = \tau_{zy} = \mu \left(\frac{\partial v}{\partial z} + \frac{\partial w}{\partial y} \right) \qquad (5.24)$$

$$\tau_{zy} = \tau_{yz} = \mu \left(\frac{\partial w}{\partial y} + \frac{\partial v}{\partial z} \right) \qquad (5.25)$$

Substituting the stress equations into the equations of motion, we have

$$\rho \left(\frac{\partial u}{\partial t} + u \frac{\partial u}{\partial x} + v \frac{\partial u}{\partial y} + w \frac{\partial u}{\partial z} \right) = -\frac{\partial P}{\partial x} + \mu \left(\frac{\partial^2 u}{\partial x^2} + \frac{\partial^2 u}{\partial y^2} + \frac{\partial^2 u}{\partial z^2} \right) + \rho g_x \qquad (5.26)$$

$$\rho \left(\frac{\partial v}{\partial t} + u \frac{\partial v}{\partial x} + v \frac{\partial v}{\partial y} + w \frac{\partial v}{\partial z} \right) = -\frac{\partial P}{\partial y} + \mu \left(\frac{\partial^2 v}{\partial x^2} + \frac{\partial^2 v}{\partial y^2} + \frac{\partial^2 v}{\partial z^2} \right) + \rho g_y \qquad (5.27)$$

$$\rho \left(\frac{\partial w}{\partial t} + u \frac{\partial w}{\partial x} + v \frac{\partial w}{\partial y} + w \frac{\partial w}{\partial z} \right) = -\frac{\partial P}{\partial z} + \mu \left(\frac{\partial^2 w}{\partial x^2} + \frac{\partial^2 w}{\partial y^2} + \frac{\partial^2 w}{\partial z^2} \right) + \rho g_z \qquad (5.28)$$

Equations 5.26 through 5.28 are called the Navier–Stokes equations. They are nonlinear and nonhomogenous partial differential equations.

5.2   FINITE ELEMENT METHOD FOR FLUID MECHANICS

The finite element method is utilized to solve the governing equations and to discretize the computational domain. Four nodes of quadrilateral elements were used for the numerical discretization. ANSYS has only this type of element for fluid dynamics. Weighted integral statements of the mass, momentum, and energy conservations over a typical element, Ω^e, are given by

$$\int_{\Omega^e} w_1 f_1 \, d\Omega = 0 \tag{5.29}$$

$$\int_{\Omega^e} w_2 f_2 \, d\Omega = 0 \tag{5.30}$$

$$\int_{\Omega^e} w_3 f_3 \, d\Omega = 0 \tag{5.31}$$

where
   $f_1, f_2,$ and f_3 are mass, momentum, and energy conservations, respectively
   $w_1, w_2,$ and w_3 are weight functions, which are equal to the interpolation functions

The choice of the weight function is restricted to the spaces of approximation functions used for pressure, velocity field, and temperature. The pressure, velocity field, and temperature can be approximated as follows:

$$P(x,t) = \sum_{l=1}^{4} \Phi_l(x), \quad P_l(t) = \Phi^{\mathrm{T}} \{P\}$$

$$u_i(x,t) = \sum_{n=1}^{4} \Psi_n(x), \quad u_i^n(t) = \Psi^{\mathrm{T}} \{u_i\} \tag{5.32}$$

$$T(x,t) = \sum_{m=1}^{4} \Theta_m(x), \quad T_m(t) = \Theta^{\mathrm{T}} \{T\}$$

where
   $\Phi, \Psi,$ and Θ are vectors of the shape functions
   $P, u_i,$ and T are vectors of nodal values for the pressure, velocity components, and
      temperature, respectively

The weight functions have the following correspondences:

$$w_1 \cong \Phi, \quad w_2 \cong \Psi, \quad w_3 \cong \Theta$$

The mass and momentum conservations can be written symbolically in the following matrix form:

$$[A^{\mathrm{T}}]\{u\} = 0 \tag{5.33}$$

$$[C]\{u\} + [K]\{u\} - [A]\{P\} = \{F\} \tag{5.34}$$

$$[D]\{T\} + [L]\{T\} = \{G\} \tag{5.35}$$

The coefficient matrices are defined by

$$A_i = \int_{\Omega^e} \frac{\partial \Psi}{\partial x_i} \Phi^{\mathrm{T}} d\Omega \tag{5.36}$$

$$C_i(u_j) = \int_{\Omega^e} \rho \Psi u_j \frac{\partial \Psi^{\mathrm{T}}}{\partial x_i} d\Omega \tag{5.37}$$

$$K_{ij} = \int_{\Omega^e} \mu \frac{\partial \Psi}{\partial x_j} \frac{\partial \Psi^{\mathrm{T}}}{\partial x_i} d\Omega \tag{5.38}$$

$$F_i = \oint_{\Gamma^e} \Psi \tau_i d\Gamma \tag{5.39}$$

$$D_i(u_j) = \int_{\Omega^e} \rho C_p (\Psi u_j) \frac{\partial \Theta^{\mathrm{T}}}{\partial x_i} d\Omega \tag{5.40}$$

$$L_{ij} = \int_{\Omega^e} k \frac{\partial \Theta}{\partial x_i} \frac{\partial \Theta}{\partial x_i}^{\mathrm{T}} d\Omega \tag{5.41}$$

$$G = \int_{\Omega^e} \Theta Q''' d\Omega + \oint_{\Gamma^e} \Theta q'' d\Gamma \tag{5.42}$$

where
   k is the thermal conductivity
   C_p is the specific heat
   τ_i is the component of the total boundary stress, which is the sum of the viscous
      boundary stress and the hydrostatic boundary stress
   q'' is the heat flux at the boundary of the elements

5.3   FLOW DEVELOPMENT IN A CHANNEL

As shown in Figure 5.2, fluid flow in a two-dimensional channel develops in the axial direction. At the inlet, the flow is uniform, and the flow Reynolds number is 100. The working fluid is water at 20°C with $\rho = 998.3\,\mathrm{kg/m^3}$ and $\mu = 1.002 \times 10^{-3}$ Pa-s.

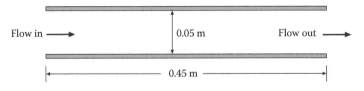

FIGURE 5.2   Channel with developing flow.

1. Show that the mass conservation principle is satisfied.
2. Determine the entrance length.

Double click on the ANSYS icon

Main Menu > Preferences

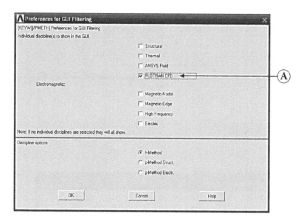

A select the FLOTRAN CFD

[OK]

This example is limited to fluid analysis. Hence, select Fluid FLOTRAN. The
2D FLOTRAN 141 element is used, and its shape is a rectangle with four nodes.
The 3D FLOTRAN 142 element is used for three-dimensional analysis. Density and
viscosity are the only properties that are needed to solve this problem.

Main Menu > Preprocessor > Element type > Add/Edit/Delete

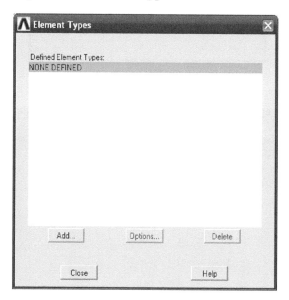

Add...

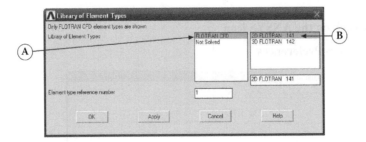

A select FLOTRAN CFD

B select 2D FLOTRAN 141

OK

Close

Main Menu > Preprocessor > Material Props > Material Models

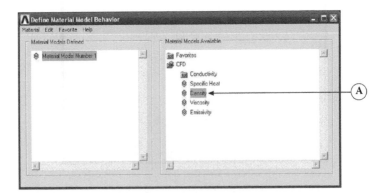

A double click on CFD > Density

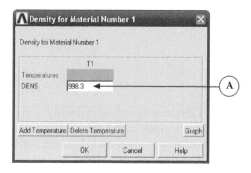

A type 998.3 in DENS

OK

Double click on CFD > Viscosity

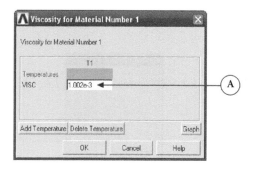

A type 1.002e-3 in VISC

OK

Close the define material model behavior window

Main Menu > Preprocessor > Modeling > Create > Areas > Rectangle > By 2 Corners

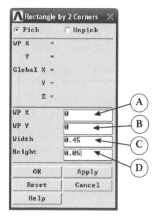

A type 0 in WP X

B type 0 in WP Y

C type 0.45 in Width

D type 0.05 in Height

OK

The smart mesh number of 1 has insufficient mesh density to give accurate results for fluid dynamics problems. The elements in the domain can be additionally increased by using the line size control in the mesh tool. Lines are divided into segments, which will be the number of elements in those lines. The lines are divided by either specifying the number of divisions or the lengths of the segments. In this example, the length of the segments is specified. The length of the segment is 0.0025 m, which means that the vertical and lateral lines will be divided into 180 and 20 segments, respectively.

Main Menu > Preprocessor > Meshing > Mesh Tool

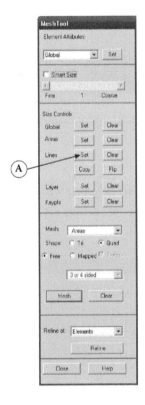

A click on Set

click on **Pick All** to select all lines, and the following window will show up.

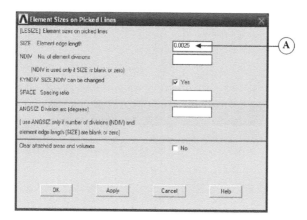

A type 0.0025

OK

A click on Mesh

Pick All

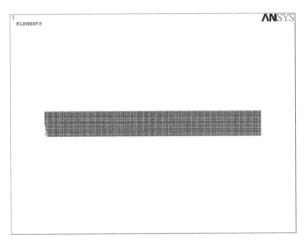

ANSYS graphics shows the mesh

Zero velocity is imposed at the upper and lower walls to simulate the wall boundary. At the inlet, the velocity is uniform. At the exit, zero pressure is imposed to simulate a free exit boundary condition. The inlet velocity can be calculated using the provided Reynolds number, as follows:

$$V = \frac{Re\mu}{\rho H} = \frac{100(1.002 \times 10^{-3})}{998.3(0.05)} = 0.002 \text{ m/s}$$

Main Menu > Solution > Define Load > Apply > Fluid/CFD > Velocity > On Lines

click at the inlet line

OK

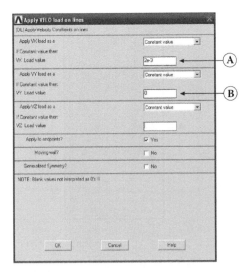

A type 2e-3 in VX Load value

B type 0 in VY Load value

OK

Main Menu > Solution > Define Load > Apply > Fluid/CFD > Velocity > On Lines

click on the two lateral lines

OK

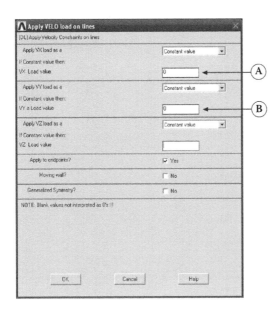

A type 0 in VX Load value

B type 0 in VY Load value

OK

Main Menu > Solution > Define Load > Apply > Fluid/CFD > Pressure DOF > On Lines

click on the exit line

OK

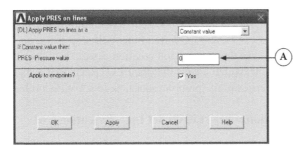

A type 0 in PRES Pressure value

OK

Main Menu > Solution > FLOTRAN Set Up > Solution Options

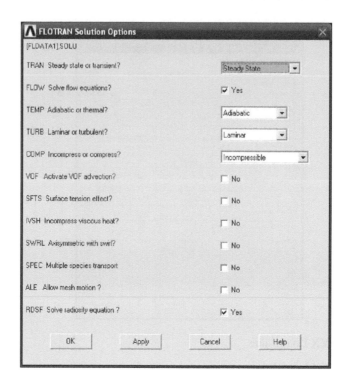

The present problem is steady state and adiabatic. Hence, keep the default setting. Notice that the ANSYS is capable of simulating turbulent and compressible flow.

OK

The maximum number of iterations is 100, and an additional 100 iterations are required if the termination criterion is not satisfied. The termination criterion for the velocity components and pressure is 1e-6. The iterations will stop if the maximum number of iterations is reached or the termination creation is satisfied. The selected material properties are MP table, which means that the FLOTRAN will use properties stored in the Material Properties in the preprocessor task to solve the problem.

Main Menu > Solution > FLOTRAN Set Up > Execution Ctrl

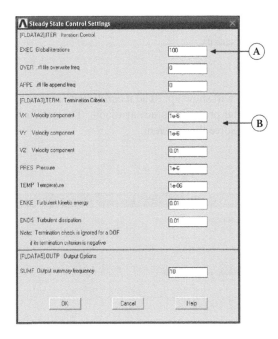

A type 100 in EXEC Global iteration

B type 1e-6 in the termination criterion for the velocity components and pressure.

OK

Main Menu > Solution > FLOTRAN Set Up > Fluid Properties

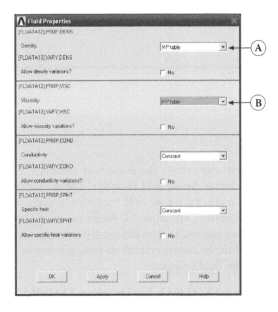

A select MP table in Density

B select MP table in Viscosity

OK

A confirmation window will show up. Read it carefully to avoid an unexpected error. The −1 indicated that the property is not available. The conductivity and specific heat are not required to solve the problem.

OK

The following command will initiate the numerical iterations. Carefully examine the normalized rate of change for all field variables. The normalized rate of change should reach termination criteria for all field variables to declare the convergence. Otherwise, additional iterations are required. When the normalized rate of change is decreasing, the solution process is approaching the convergence. If the solution is diverged, either some of the boundary conditions were missing or incorrect, or there were wrong inputs for the fluid properties.

Main Menu > Solution > FLOTRAN Set Up > Run FLOTRAN

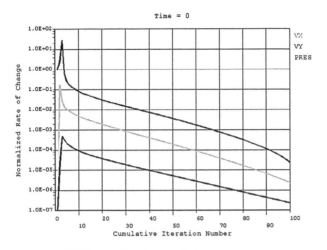

ANSYS graphics shows solution convergence

OK

As shown in the normalized rate of change, the solution reaches the maximum number of iterations, which is 100, without reaching the termination criterion, which is 1e-6. Therefore, additional iterations are required to solve the problem.

Main Menu > Solution > FLOTRAN Set Up > Run FLOTRAN

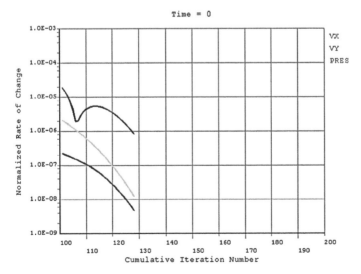

ANSYS graphics shows solution convergence

OK

Now, the normalized rate of change reaches the termination criterion. Hence, convergence is attained. The Last Set is selected in the Read Results to ensure that the data from the last iteration set is loaded for the postprocessor. Otherwise, no results will be shown in the postprocessor. A plot of velocity vectors is presented in the postprocessor task, followed by that of a velocity profile along the centerline of the channel and at the exit.

Main Menu > General Postproc > Read Results > Last Set

Main Menu > General Postproc > Plot Results > vector Plot > Predefined

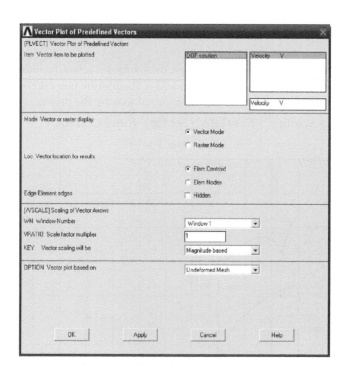

OK

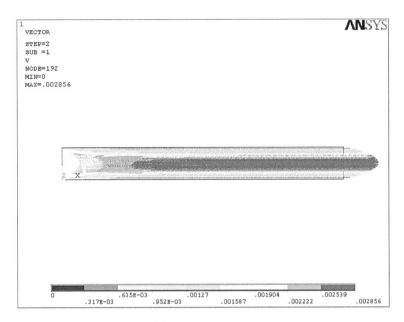

ANSYS graphics shows vector for the velocity

The red arrows are for the maximum velocity in the channel, while the blue arrows are for minimum velocity. The velocity is maxima at the centerline, and zero at the wall. The developing region is clearly shown at the entrance region of the channel. Notice that the velocity has a parabolic velocity profile at the exit of the channel. To determine the entrance length, a plot of the x-velocity component along the channel's centerline is created using the path operation in the postprocessor. The x-velocity component should be increased from its initial value at the inlet until it becomes constant. The path is created by specifying two points, one at the inlet and the other at the exit along the centerline. The number of division is the number of points used to create the plot. A higher number of divisions will create a smother plot.

Main Menu > General Postproc > Path Operation > Define Path > By Location

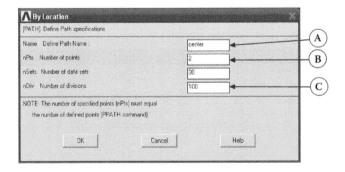

A type center in Define Path Name. The name of the path is optional

B type 2 in Number of points

C type 100 in Number of divisions

OK

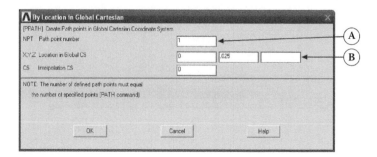

A type 1 in Path point number

B type 0 and 0.025 in Location in Global CS

OK

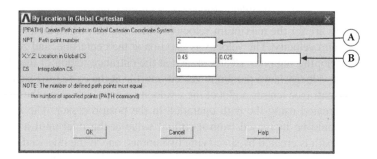

A type 2 in Path point number

B type 0.45 and 0.025 in Location in Global CS

OK

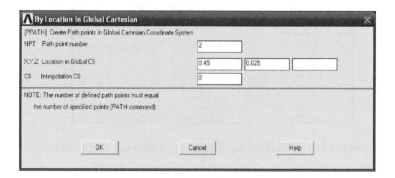

Cancel

Main Menu > General Postproc > Path Operation > Map onto Path

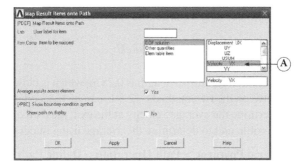

A select Velocity VX

Now, the stored variable, VX, is ready to be plotted. In the Plot path item, there are two options. The stored data can be either plotted or listed. The list results can be exported to other graphical software such as EXCEL.

Main Menu > General Postproc > Path Operation > Plot path Item > On Graph

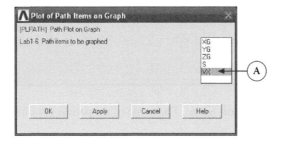

A select VX

OK

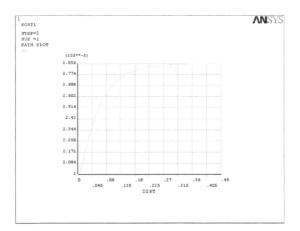

ANSYS graphics shows x-velocity profile along the centerline of the channel

The above figure indicates that the flow is fully developed at the exit of the channel. The x-velocity component becomes invariant at a distance of 0.315 m from the entrance. Experimentally, the entrance length can be determined using the following equation:

$$L = 0.056 ReH$$

Using the above equation, the entrance length is 0.28 m, which is close to the ANSYS results. The error between the two methods, 12.5%, can be reduced if a finer mesh is used. In order to determine the average velocity at the exit, the path operation is used to plot the velocity profile and to determine the average using integration.

Main Menu > General Postproc > Path Operation > Define Path > By Location

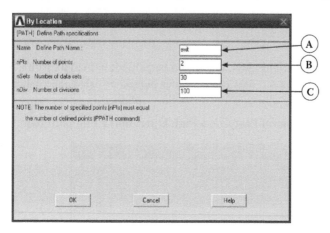

A type exit in Define Path Name. The name of the path is optional

B type 2 in Number of points

C type 100 in Number of divisions

$\boxed{\text{OK}}$

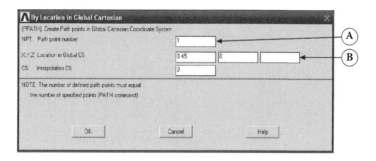

A type 1 in Path point number

B type 0.45 and 0 in Location in Global CS

OK

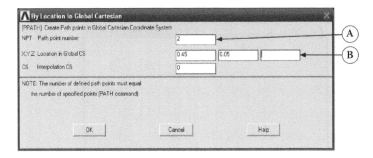

A type 2 in Path point number

B type 0.45 and 0.05 in Location in Global CS

OK

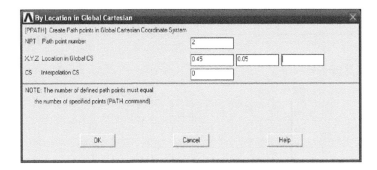

Cancel

Main Menu > General Postproc > Path Operation > Map onto Path

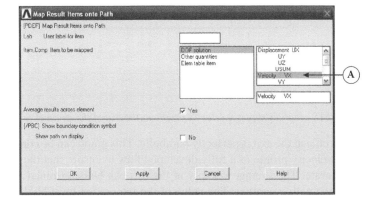

A select Velocity VX

OK

Main Menu > General Postproc > Path Operation > Plot path Item > On Graph

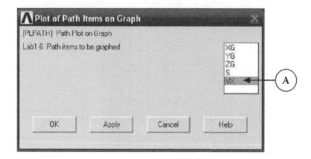

A select VX

OK

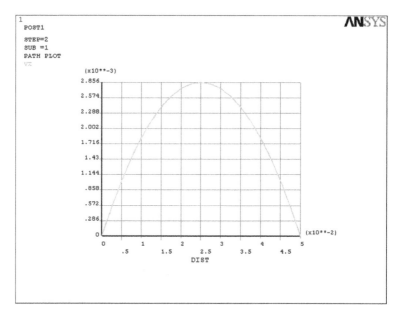

ANSYS graphics shows x-velocity profile at the exit of the channel

The velocity profile at the exit is perfectly parabolic. This graph can be compared to the analytical velocity profile of a fully developed flow to ensure that the obtained solution is accurate. The average velocity at the exit can be determined using the integration in the path operation. The value of the integration must be divided by the path length to get the average value of the variable. The number 20 in the Factor is the inverse of the path length. Selecting S in Lab2 means that the integration is performed along the path.

Main Menu > General Postproc > Path Operation > Integrate

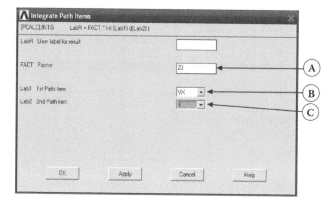

A type 20 in Factor

B select VX in 1st Path item

C select S in 2nd Path item

OK

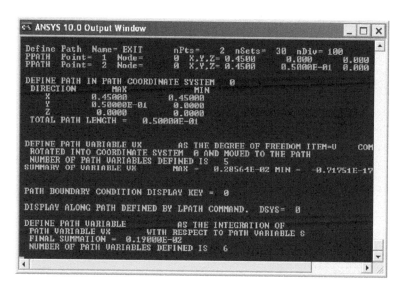

The ANSYS Output window shows the value of the integration, which is 0.19e-2. This is almost equal to the inlet velocity with an error of 5%.

5.4 FLOW AROUND A TUBE IN A CHANNEL

For the heat exchanger tube shown in Figure 5.3, determine the pressure drop in the channel and the maximum shear stress at the surface of the tube. The working

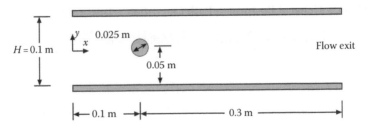

FIGURE 5.3   Flow around a tube in a channel.

fluid is air, and the inlet velocity is fully developed with an average velocity of $U_m =$ 0.01 m/s. At the exit, atmospheric pressure is applied. Use the following equation for the velocity profile at the inlet:

$$u(y) = \frac{3}{2} U_m \left[1 - \left(\frac{2y}{H} \right)^2 \right]$$

Double click on the ANSYS icon

Main Menu > Preferences

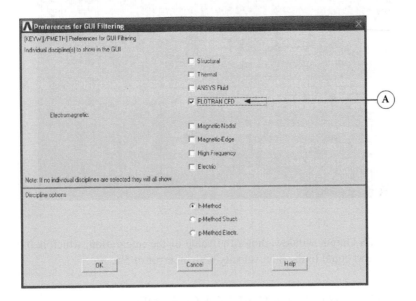

A select the FLOTRAN CFD

Main Menu > Preprocessor > Element type > Add/Edit/Delete

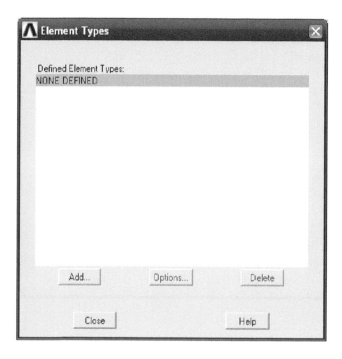

Add...

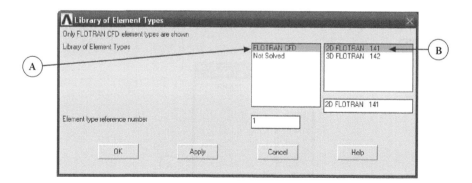

A select FLOTRAN CFD

B select 2D FLOTRAN 141

OK

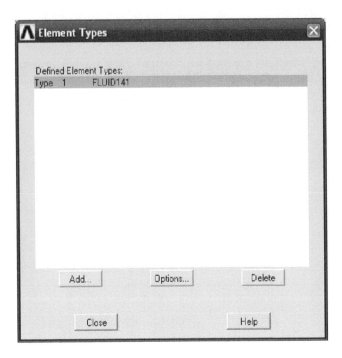

Since this problem is fluid dynamics, density and viscosity are required to solve it. Air in the material library of the ANSYS is used, and this is done in the FLOTRAN setup in the solution task. The geometry is modeled by creating a square and a circle. A Boolean operation is utilized to remove the circle from the square using the overlap and delete commands. Alternatively, the circle can directly be removed using the subtract command.

Main Menu > Preprocessor > Modeling > Create > Areas > Rectangle > By 2 Corners

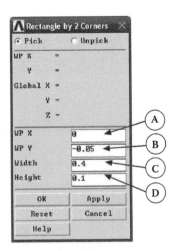

A type 0 in WP X

B type −0.05 in WP Y

C type 0.4 in Width

D type 0.1 in Height

OK

Main Menu > Preprocessor > Modeling > Create > Areas > Circle > Solid Circle

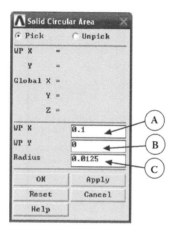

A type 0.1 in WP X

B type 0 in WP Y

C type 0.0125 in Radius

OK

Main Menu > Preprocessor > Modeling > Operate > Booleans > Overlap > Areas

Pick All

Main Menu > Preprocessor > Modeling > Delete > Area Only

click on the circle to highlight it

OK

Lines in the created geometry are divided into segments, and the number of segments will be the number of elements in those lines. Increasing the number of segments will create a finer mesh. Line division can be accomplished by specifying either the number of segments or the length of the segments. In this example, the length of the segments is specified. The length of the segment is 0.0025 m.

Main Menu > Preprocessor > Meshing > Mesh Tool

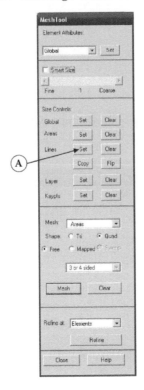

A click on Set

Pick All

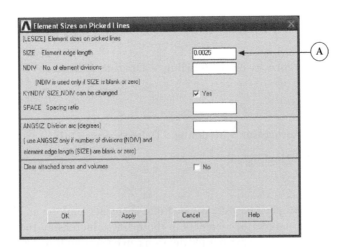

A type 0.0025 in Element edge length

OK

A mesh

Pick All

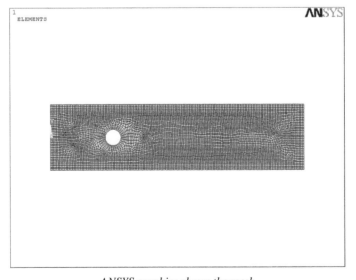

ANSYS graphics shows the mesh

The modeling and meshing tasks are now completed. The hydrodynamics boundary conditions are applied. At the inlet, Vx is a function of the y-direction and Vy is 0. At the exit, zero pressure is imposed to simulate a free exit boundary condition. Zero velocity components are applied at the surface of the cylinder and lateral walls to simulate a wall boundary condition.

The inlet velocity profile can be applied easily with the function editor feature of the ANSYS. ANSYS will create a data table from the velocity function, and apply it to the selected lines. To ensure that the equation is written correctly, a graph of velocity is created in the ANSYS graphics.

Main Menu > Solution > Define Load > Apply > Functions > Define/Edit

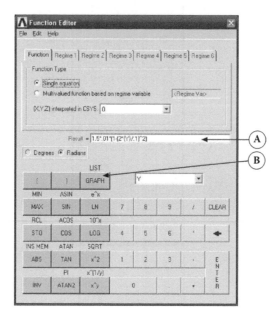

A Type the equation: 1.5*0.01*(1−(2*{Y}/.1)^2)

B click on GRAPH

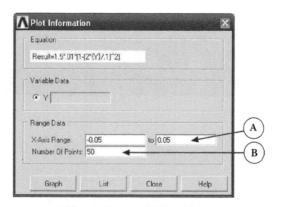

A type −0.05 and 0.05 for X-Axis Range

B type 50 in Number Of Points

| Graph |

| Close |

The range of the data in the x-axis are −0.05 and 0.05, while 50 is the number of points to be plotted. Number Of Points has nothing to do with the accuracy of the results, and a higher number of it will just create a smother plot.

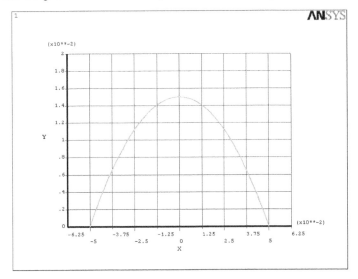

ANSYS graphics shows the inlet velocity profile

The above figure indicates that the velocity profile equation is correctly presented in the ANSYS. The velocity profile equation must be saved.

In the Equation Editor click on File then Save

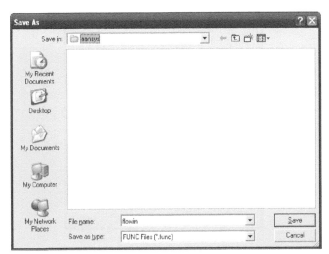

Save the file as flowin. The file name is optional.

Close the Function editor window

After saving the function, you must load it to the ANSYS solution using the read file in function.

Main Menu > Solution > Define Load > Apply > Functions > Read File

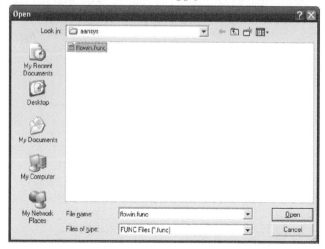

A select flowin.func

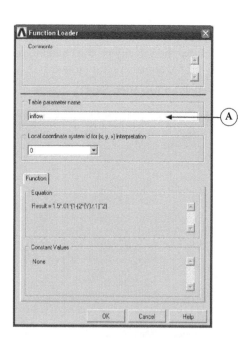

A type inflow in Table parameter name. The name of the table is optional. The "inflow" will be shown later when the boundary conditions are applied.

OK

Main Menu > Solution > Define Load > Apply > Fluid/CFD > Velocity > On Lines

click on the lines of the upper and lower walls of the channel, and the surface of the cylinder

OK

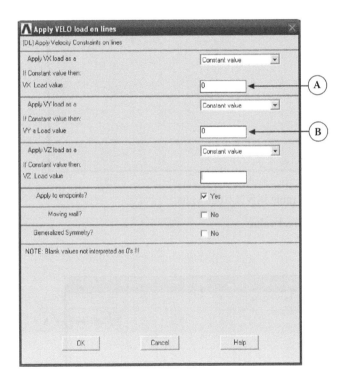

A type 0 in VX Load value

B type 0 in VY Load value

OK

Main Menu > Solution > Define Load > Apply > Fluid/CFD > Velocity > On Lines

click on the channel entrance

OK

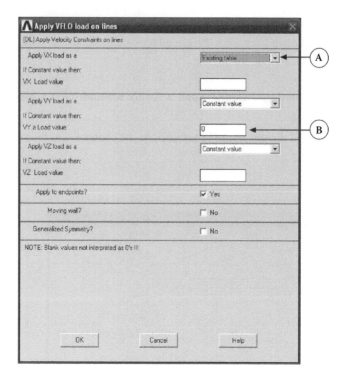

A select Excising table in VX Load value

B type 0 in VY Load value

OK

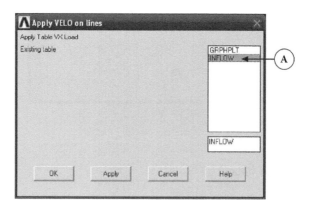

A select INFLOW

OK

Main Menu > Solution > Define Load > Apply > Fluid/CFD > Pressure DOF > On Lines

click on the channel's exit line

OK

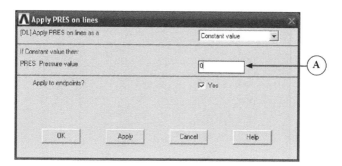

A type 0 in PRES Pressure value

OK

Main Menu > Solution > FLOTRAN Set Up > Solution Options

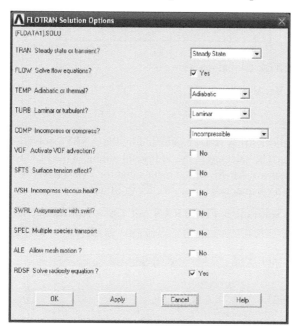

The present problem is steady state and adiabatic. Hence, keep the default setting.

OK

The maximum number of iterations is 100, and an additional 100 iterations are required if the termination criterion is not satisfied. The termination criterion for the velocity components and pressure is 1e-6. The iterations will stop if the maximum number of iterations is reached or the termination criterion is satisfied for all field variables.

Main Menu > Solution > FLOTRAN Set Up > Execution Ctrl

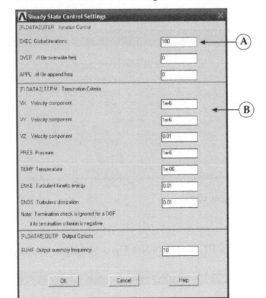

A type 100 in EXEC Global iteration

B type 1e-6 for Vx, Vy, and PRES

OK

By default, the shear stress is not calculated by the ANSYS solver. To calculate the shear stress, it must be selected in the additional output in the FLOTRAN setup. The material properties are Air-SI, which means that the FLOTRAN will use properties stored in the ANSYS material library for air in SI units.

Main Menu > Solution > FLOTRAN Set Up > Additional Out > RFL Out Derived

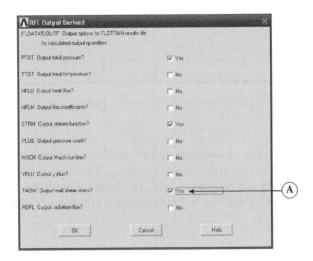

A select TAUW output wall shear stress

OK

Main Menu > Solution > FLOTRAN Set Up > Fluid Properties

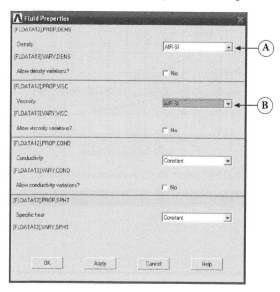

A select AIR-SI in Density

B select AIR-SI in Viscosity

OK

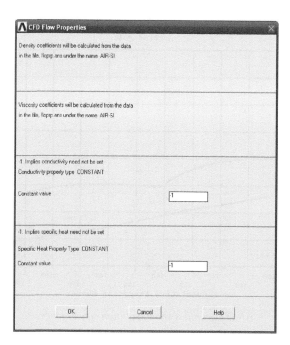

OK

Main Menu > Solution > FLOTRAN Set Up > Run FLOTRAN

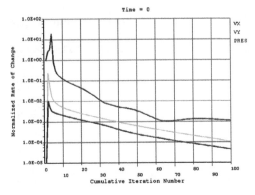

ANSYS graphics shows solution convergence

OK

As shown in the normalized rate of change, the solution reaches the maximum number of iterations, which is 100, without reaching the termination criterion, which is 1e-6. Therefore, additional iterations are required to solve the problem.

Main Menu > Solution > FLOTRAN Set Up > Run FLOTRAN

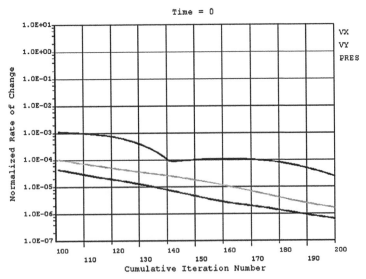

ANSYS graphics shows solution convergence

OK

The solution is still not converged. Therefore, an additional iteration is required to solve the problem.

Main Menu > Solution > FLOTRAN Set Up > Run FLOTRAN

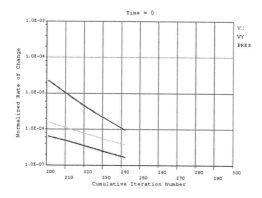

ANSYS graphics shows solution convergence

OK

Now, the normalized rate of change reaches the termination criterion. Hence, the convergence is reached. The Last Set is selected in the Read Results to ensure that the data from the last iteration set is loaded for the postprocessor. Otherwise, no results will be shown in the postprocessor. A plot of velocity vectors is presented in the postprocessor task, followed by pressure drop calculations.

Main Menu > General Postproc > Read Results > Last Set

Main Menu > General Postproc > Plot Results > vector Plot > Predefined

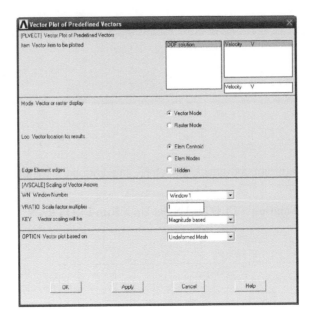

OK

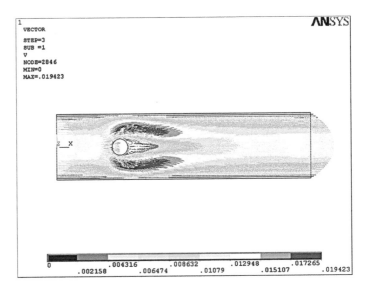

ANSYS graphics shows vector for the velocity

The red arrows are for the maximum velocity in the channel, while the blue arrows
are for minimum velocity. The wake flow behind the cylinder is clearly visible in the
figure, and the flow is very slow at this region. The flow velocity is maxima above
and below the cylinder. Notice that the velocity has a parabolic profile at the inlet and
exit of the channel. To calculate the pressure drop, the average pressure at the inlet
is determined using the path operation in the postprocessor. The path at the inlet is

created by specifying two points. The number of divisions is the number of points used to create the plot, and using a higher number of divisions will create a smoother plot.

Main Menu > General Postproc > Path Operation > Define Path > By Location

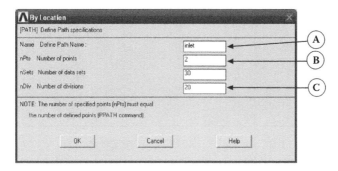

A type inlet in Define Path Name. The name of the path is optional

B type 2 in Number of points

C type 20 in Number of divisions

OK

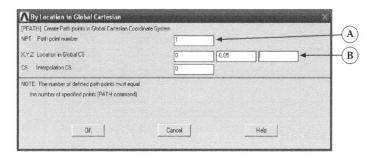

A type 1 in Path point number

B type 0 and −0.05 in Location in Global CS

OK

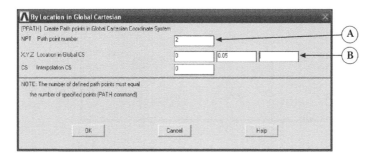

A type 2 in Path point number

B type 0 and 0.05 in Location in Global CS

OK

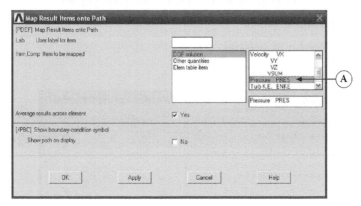

Cancel

Main Menu > General Postproc > Path Operation > Map onto Path

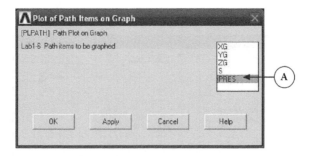

A select Pressure PRES

OK

Main Menu > General Postproc > Path Operation > Plot path Item > On Graph

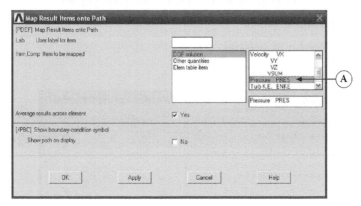

A select PRES

OK

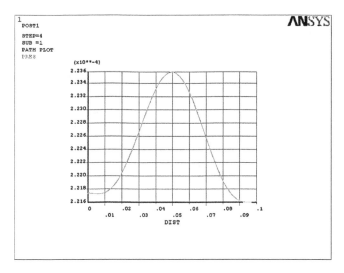

ANSYS graphics shows pressure distribution at the inlet of the channel

The average pressure at the inlet can be determined using the integration in the path operation. The value of the integration must be divided by the path length to get the average value of the variable. The number 10 in the Factor is the inverse of the path length. Selecting S in Lab2 means that the integration is performed along the path.

Main Menu > General Postproc > Path Operation > Integrate

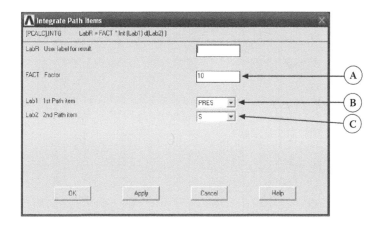

A type 10 in Factor

B select PRES in 1st Path item

C select S in 2nd Path item

OK

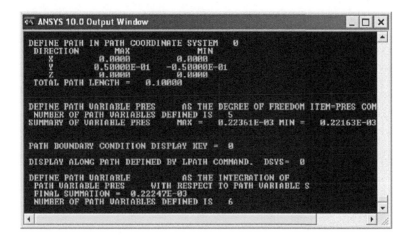

The ANSYS Output window shows the average pressure at the inlet, which is $0.22247\text{e-}3\,\text{N/m}^2$. At the exit, the pressure is specified as a boundary condition and is equal to zero. Hence, the pressure drop in the channel, is equal to $0.22247\text{e-}3\,\text{N/m}^2$.

To determine the maximum shear stress at the surface of the cylinder, the distribution of the shear stress using the path operation is employed. A path is created around the cylinder using the defined path by nodes. It is necessary to zoom the cylinder's region for better node selection.

ANSYS Utility Menu > PlotCtrls > Pan Zoom Rotate …

click on Box Zoom, and make a box around the cylinder, as shown below.

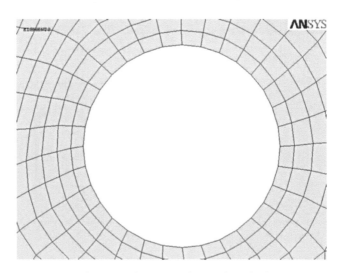

Mesh view at the region close to the cylinder

Using the mouse, carefully select all nodes in the surface of the cylinder, and the selection should be in order. Double selections or backward selections will produce a wrong result. Start the selection from the node at the front stagnation point of the cylinder.

Main Menu > General Postproc > Path Operation > Define Path > By Nodes

click on nodes on the surface of the cylinder

OK

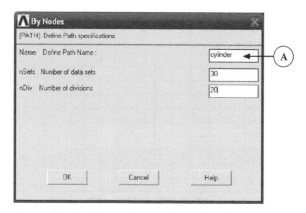

A name the path as cylinder. The path name is optional

OK

Main Menu > General Postproc > Path Operation > Map onto Path

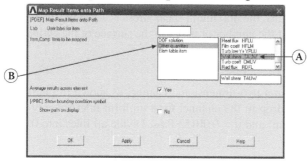

A select Other Quantities

B select wall shear TAUW

OK

Main Menu > General Postproc > Path Operation > Plot path Item > On Graph

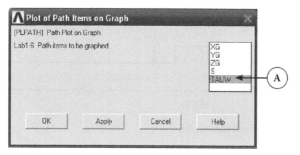

A select TAUW

OK

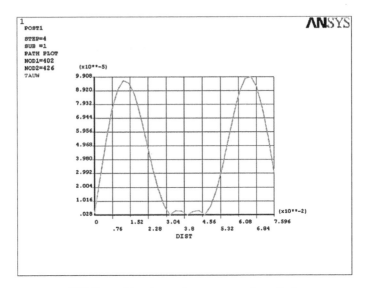

ANSYS graphics shows shear stress at the cylinder

For the above results, the maximum shear stress is equal to 9.908e-5 N/m². The plot
can be smoother if the number of nodes at the surface of the cylinder is increased.

PROBLEMS

5.1 Water flows in a two-dimensional channel as shown in Figure 5.4. Determine
the pressure drop in the channel for a uniform inlet velocity of 0.001 m/s. Let
$\rho = 998.3$ kg/m³ and $\mu = 0.00152$ Pa-s. Divide all lines into 0.0025 m segments
for meshing.

5.2 Air flows in a two-dimensional channel as shown in Figure 5.5. Determine the
maximum shear stress at the cylinder wall for Reynolds numbers of 100 and
200. Let $\rho = 1.204$ kg/m³ and $\mu = 1.82$e-5 Pa-s. Divide all lines into 0.01 m segments for meshing.

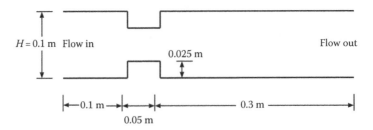

FIGURE 5.4 Channel for Problem 5.1.

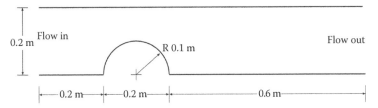

FIGURE 5.5 Channel for Problem 5.2.

5.3 The two-dimensional channel shown in Figure 5.6, contains a triangular object used to simulate an airfoil. The inlet flow is uniform and it is 0.01 m/s, and the working fluid is air. Divide all lines into 0.0075 m segments for meshing. Determine

1. Maximum velocity in the x-direction
2. Maximum shear stress on the object
3. Pressure drop in the channel

5.4 A heat exchanger consists of a channel with four cylinders with a diameter of 0.025 m, as shown in Figure 5.7. The working fluid is air, entering from the left

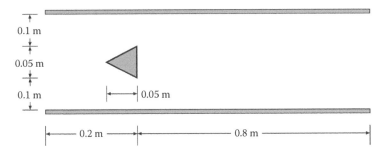

FIGURE 5.6 Channel containing a triangular object for Problem 5.3.

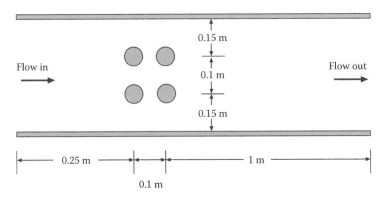

FIGURE 5.7 Channel with four cylinders for Problem 5.4.

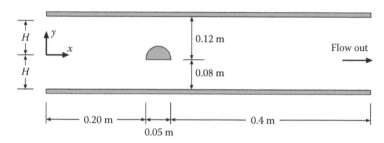

FIGURE 5.8   Channel with a half cylinder object for Problem 5.5.

side with an average velocity of 0.005 m/s. The radius of the cylinders is 0.025 m. Determine the maximum velocity and pressure drop in the channel.

5.5 The two-dimensional channel shown in Figure 5.8, contains a half cylinder object used to simulate an airfoil. Determine the average shear stress at the upper surface of the object and verify that the mass conservation is satisfied. The working fluid is air. The average inlet velocity is 0.005 m/s, and the inlet velocity has the following velocity profile:

$$u(y) = \frac{3}{2}U_m\left[1 - \left(\frac{2y}{H}\right)^2\right]$$

6 Multiphysics

6.1 INTRODUCTION

In practice, the engineering problems that are to be simulated are multiphysics, but there has not been sufficient capability in the past to address the full physics of the problems. Therefore, there has been a tendency to simplify the problems by either focusing on the primary physics or decoupling the physics. For example, for the heat exchanger analyses, only fluid flow is solved and then followed by the thermal analysis. Recently, significant improvements have been achieved in both hardware and software capabilities that made the multiphysics simulations inexpensive and accurate. As computing power increases, the ability to model the full physics is becoming a practical possibility. Hence, there is an increasing demand for multiphysics simulation software. ANSYS has the capability to model such complicated simulations. ANSYS can effectively simulate the thermal–structural, thermal–fluid, and fluid–structural problems.

In this chapter, setting up of problems involving multiphysics is introduced. The level of coupling for multiphysics simulations is illustrated. Examples of thermal–structure, thermal–fluid, and fluid–structure systems are presented. The coupling between the physical phenomena can be classified into three levels: low, medium, and high. With a low level, it may be sufficient to use a simple one-way coupling with file transfer between two physics, as analysis in thermal–structure and thermal–fluid with temperature-dependent properties. In the thermal–structure, the thermal analysis is done first, and then the temperature distribution is transferred to the structural analysis as a boundary condition. This analysis is fast and the thermal and structural analysis is completely separated, because the structural deformation has a negligible effect on the temperature distribution. The process is shown in Figure 6.1.

The problem with a medium level of coupling, the solution of the each physics depends on other, but requires no mesh movement, such as thermal–fluid analysis with temperature-dependent properties. In the thermal–fluid analysis, the properties of the fluid, such as the density, are a function of temperature. In this analysis, the fluid is solved first, followed by thermal analysis. The properties of the fluid are updated, and then the fluid is solved again. This process is continued until the solution is fully converged. The process is shown in Figure 6.2. If all properties of the fluid are independent of the temperature, the heat transfer and fluid flow can be solved separately, and the simulation is considered as a low-level analysis.

The fluid–structural analysis is considered as high level. It requires a degree of compatibility in the solver technologies and often involves a mesh movement, which may be accounted for by using of the arbitrary Lagrangian–Eulerian (ALE) method. The computation time for this analysis is high, and it requires high-performance

FIGURE 6.1   Low-level analysis.

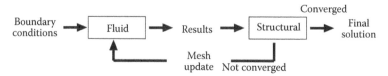

FIGURE 6.2   Medium-level analysis.

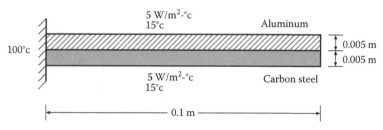

FIGURE 6.3   High-level analysis.

computers. The process is shown in Figure 6.3. There are two types of fluid–structural analysis. In the first type, the solid is moving in a specified way and the fluid pressure has no effect on the movement of the solid. In the second type, which is much more complex, the fluid pressure affects the movement of the structure, and possible structural deformation is occurred.

6.2   THERMAL AND STRUCTURAL ANALYSIS OF A THERMOCOUPLE

High-temperature furnaces use thermocouples to control their temperature. Figure 6.4 shows the thermocouple. It consists of aluminum and carbon steel plates attached

FIGURE 6.4   Thermocouple.

to each other. The left end is fixed and maintained at 100°C, and the thermocouple is exposed to a free convection boundary condition with $h = 5\,W/m^2\text{-}°C$ and $T_\infty = 15°C$. Display the y-displacement contours, and determine the maximum displacement in the y-direction. The thermophysical properties of the aluminum and carbon steel are shown in the following table:

Property	Aluminum	Carbon Steel
Thermal conductivity (W/m-K)	83	111
Young modulus (Pa)	70×10^9	210×10^9
Poisson's ratio	0.33	0.29
Thermal expansion (1/K)	23×10^{-6}	12×10^{-6}

This problem is considered as multiphysics and physics are not coupled because the structure and heat transfer are solved separately. The deformation of the thermo-couple has no effect on the heat transfer. The heat transfer must be solved first, and then the structure.

Double click on the ANSYS icon

Main Menu > Preferences

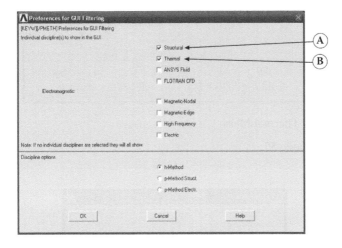

A select the Structure

B select the Thermal

A solid thermal element is used, and its shape is quadratic with four nodes. This ele-ment will be replaced by structural elements when the thermal part is solved.

Main Menu > Preprocessor > Element type > Add/Edit/Delete

Add...

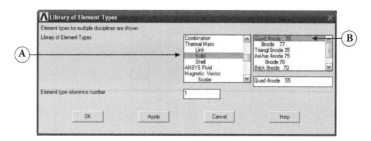

A select Solid in Thermal Mass

B select Quad 4node 55

OK

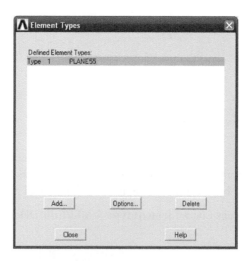

Close

For the material properties, the modulus of elasticity and Poisson's ratio are required to solve the problem. For the thermal part, the thermal conductivity is required. Additionally, the thermal expansion for both materials is very important to account for the deflection of the thermocouple. There will be different sets of properties for aluminum and carbon steel. The Material number 1 in the Material Models Model Behavior is for aluminum, while material number 2 is for carbon steel.

Main Menu > Preprocessor > Material Props > Material Models

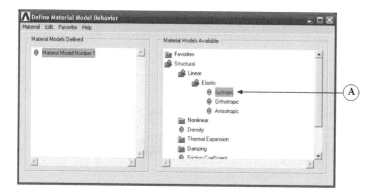

A Double click on Structural > Linear > Elastic > Isotropic

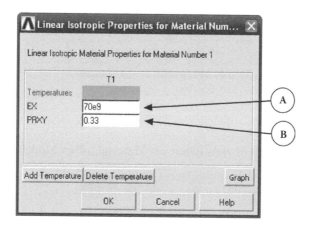

A type 70e9 in EX

B type 0.33 in the PRXY

OK

Double click on Structure > Thermal Expansion > Secant Coefficient > Isentropic

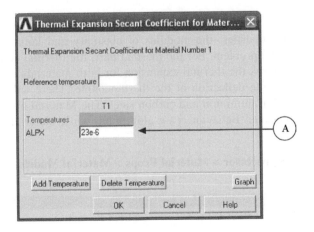

A type 23e-6

OK

Double click on Thermal > Conductivity > Isotropic

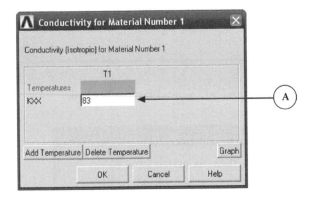

A type 83 in KXX

OK

In the Define Material Models Behavior: Material > New Model

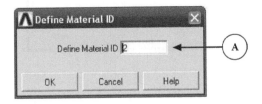

A type 2 in Define Material ID

OK

Select material number 2 in the Material Model Defined, then

Double click on Structure > Linear > Elastic > Isotropic

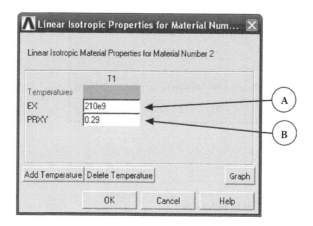

A type 210e9 in EX

B type 0.29 in the PRXY

OK

Double click on Structure > Thermal Expansion > Secant Coefficient > Isentropic

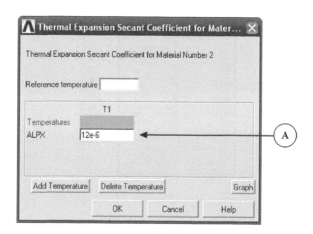

A type 12e-6

OK

Double click on Thermal > Conductivity > Isotropic

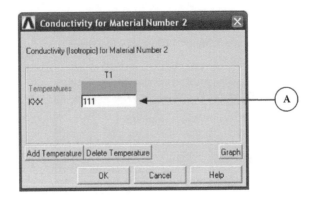

A type 111 in KXX

OK

Close the material models behavior

The geometry of the problem is relatively simple. Two rectangles are created and then they are glued.

Main Menu > Preprocessor > Modeling > Create > Areas > Rectangle > By 2 Corners

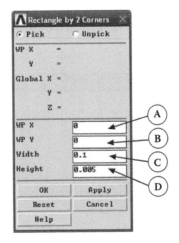

A type 0 in WP X

B type 0 in WP Y

C type 0.1 in Width

D type 0.005 in the Height

Apply

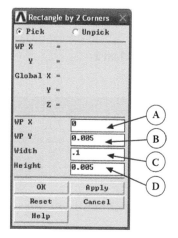

A type 0 in WP X

B type 0.005 in WP Y

C type 0.1 in Width

D type 0.005 in the Height

OK

Main Menu > Preprocessor > Modeling > Operate > Booleans > Glue > Areas

Pick All

Main Menu > Preprocessor > Meshing > Mesh Tool

A select Areas

B click on Set

Using mouse, select the carbon steel area

OK

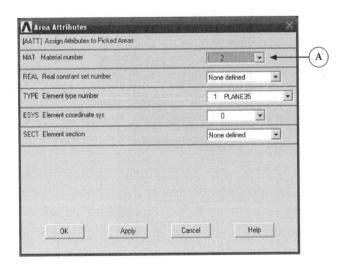

A select 2 in Material number

OK

Utility Menu > PlotCtrls > Numbering…

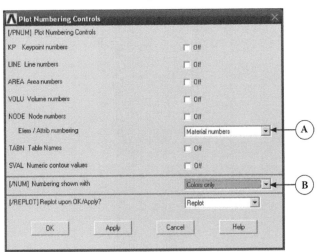

A select Material numbers

B select Colors only

OK

Main Menu > Preprocessor > Meshing > Mesh Tool

A select Smart Size

B set the level to 1

C Mesh

Click on Pick All

Close

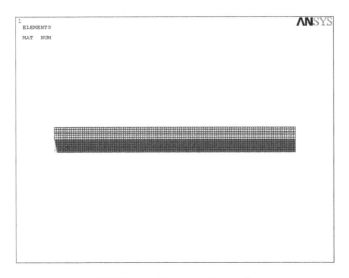

ANSYS graphics shows the mesh

Main Menu > Solution > Define Load > Apply > Thermal > Temperature > On Lines

Click on the left boundary where temperature boundary condition is applied.

OK

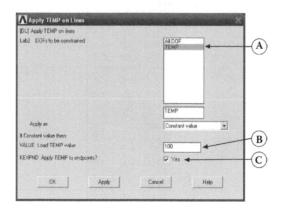

A select TEMP

B type 100 in Value

C select Yes

OK

Main Menu > Solution > Define Load > Apply > Thermal > Convection > On Lines

Click on the surfaces where convection boundary condition is applied. Do not click on the surface between the two materials.

OK

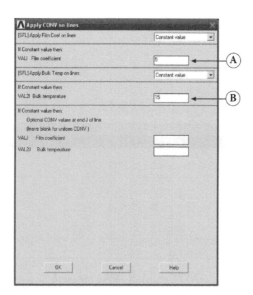

A type 5 in Film coefficient

B type 15 in Bulk temperature

OK

Now, the problem is ready to be solved as a heat transfer problem.

Main Menu > Solution > Solve > Current LS

OK

Close

The thermal solution is now completed. The thermal elements must be replaced by structural elements with the same element type and distribution. This can be done using the Switch element type in the preprocessor. The temperature contours is plotted to check if the thermal boundary conditions are correctly applied.

Main Menu > General Postproc > Plot Results > Contour Plot > Nodal Solu

A click Nodal Solution > DOF Solution > Nodal Temperature

OK

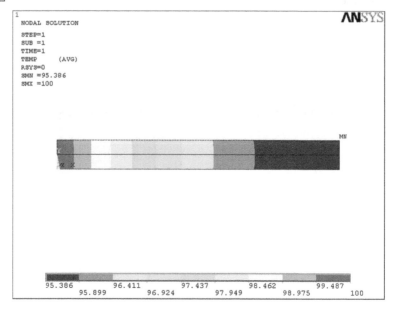

ANSYS graphics shows the temperature contours

The temperature contours shows the left side is maintained at 100°C, and the temperature is decreasing in the lateral direction due to convective cooling. In general, the temperature contour is as expected. The following steps are very important: the element type is switched to structural type, and the nodal temperature solution from the thermal analysis is loaded into the structural element nodes.

Main Menu > Preprocessor > Element type > Switch Elem Type

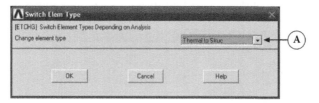

A select Thermal to Struc

OK

Close

Main Menu > Solution > Define Load > Apply > Structure > Temperature > form therm Analy

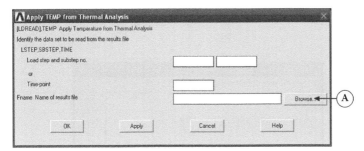

A click on Browse

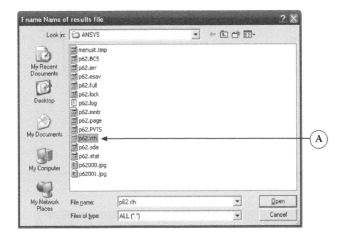

A Select from the windows the thermal results file that has an extension of rth

Open

OK

The structural boundary condition is the zero displacement at the left side of the thermocouple.

Main Menu > Solution > Define Load > Apply > Structural > Displacement > On Lines

In the ANSYS graphics, click on the left lines where displacement is applied

OK

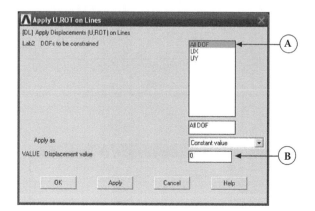

A select All DOF

B type VALUE to 0

OK

Main Menu > Solution > Solve > Current LS

OK

OK

Main Menu > General Postproc > Plot Results > Deformed Shape

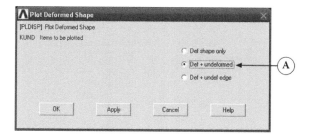

A select Def + undeformed

OK

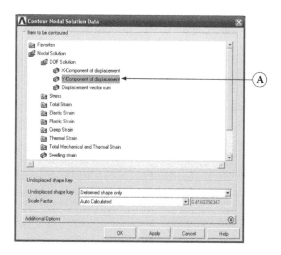

ANSYS graphics shows the thermocouple before and after deformation

The plot shows the deflection of the thermocouple because the thermal expansion of aluminum is higher than copper.

Main Menu > General Postproc > Plot Results > Contour Plot > Nodal Solu

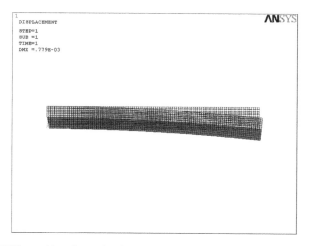

A select Nodal Solution > DOF Solution > Y-Component of displacement

OK

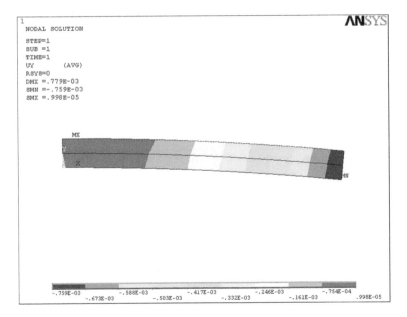

ANSYS graphics shows displacement in the y-direction

The ANSYS results indicate that the maximum displacement in the y-direction is equal to 0.759×10^{-3} m.

6.3   CHIPS COOLING IN A FORCED CONVECTION DOMAIN

Predict the thermal characteristics of two electronic chips mounted on a channel using the ANSYS. The configuration is shown in Figure 6.5. The working fluid is air, and thermophysical properties of the chips are listed in table below, and for air, use ANSYS properties. The inlet velocity is 0.01 m/s, while the exit pressure is the atmospheric. The inlet temperature is 25°C, and 10 W is generated in each chip. Consider the problem as a steady heat transfer. Determine

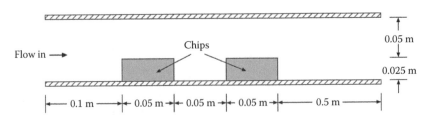

FIGURE 6.5   A channel with two chips.

1. Maximum velocity in the x-direction
2. Maximum temperature in the domain
3. Average heat transfer coefficient at the surface of the first chip

Property of the Chips

Density	$2500\,kg/m^3$
Conductivity	$0.25\,W/m\text{-}°C$
Specific heat	$850\,J/kg\text{-}°C$

Electronic manufacturers are continuously providing the market with high-performance devices, but with high heat dissipation. Heat highly affects the performance and durability of electronic devices more than any other factors, and operating electronic devices at a temperature higher than its recommended operating value will significantly affect their reliability and functionality. Thermal analysis of a channel containing multiple heated blocks subjected to forced convection flow is extensively addressed in the literature because it simulates integrated circuit chips placed on a horizontal board. In these analyses, researchers are focusing on the temperature distribution within the chips, the maximum temperature of the chips, and distributions of the local Nusselt number along the chips' surface. Temperature distribution within the chips is typically used to predict the reliability of some of its components, and to establish a guide for safe operating conditions. The Nusselt number is used to estimate the heat flow out of the chips, and to determine the required cooling load. In addition, researchers are interested in the friction factors on the surface of the chips, and the pressure drop in the channel, which are used to measure the required fan power.

This problem is considered as multiphysics, and physics are not coupled because the fluid flow and heat transfer are solved separately. In this problem, the ANSYS will solve the fluid flow first, and then it uses the velocity components to determine the temperature. ANSYS Flotran CFD can solve fluid dynamics problems, as well as the thermal problems. The selected FLOTRAN 141 element has also temperature as a degree of freedom besides the velocity components and pressure.

Double click on the ANSYS icon

Main Menu > Preferences

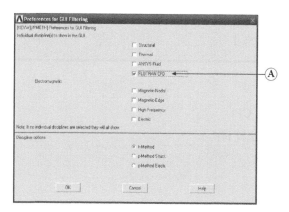

A select the FLOTRAN CFD

OK

Main Menu > Preprocessor > Element type > Add/Edit/Delete

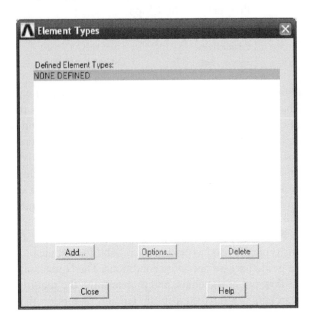

Add...

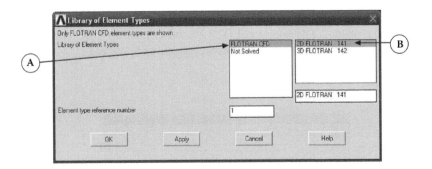

A select FLOTRAN CFD

B select 2D FLOTRAN 141

OK

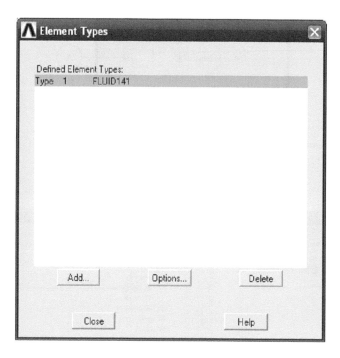

Close

There will be different sets of properties for air and chips. The Material number 1 in the Material Models Model Behavior is reserved for the fluid only, and material number 2 to 10 is for the solids. For air, Air-SI in the Flotran materials library will be used.

Main Menu > Preprocessor > Material Props > Material Models

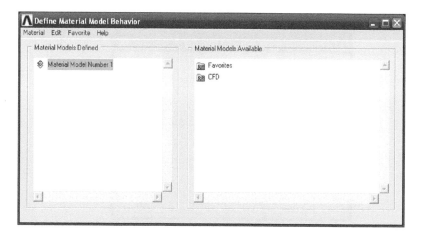

In the Define Material Models Behavior menu: Material > New Model

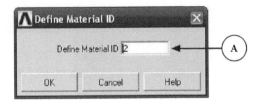

A type 2 in Define Material ID

OK

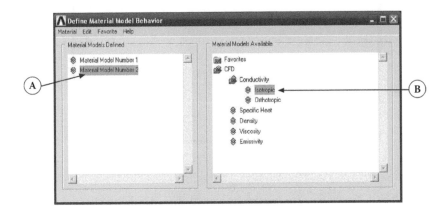

A Select Material Model Number 2

B double click on CFD > Conductivity > Isotropic

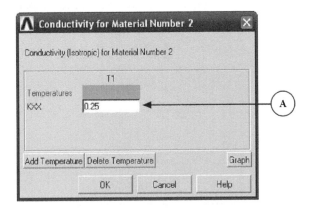

A type 0.25 in KXX

OK

Double click on CFD > Specific Heat

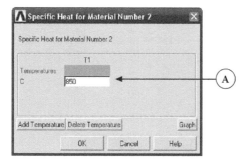

A type 850 in C

OK

Double click on CFD > Density

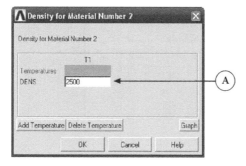

A type 2500 in DENS

OK

Close define material models behavior

The geometry is created by first creating a large rectangular area for fluid flow. Then, the two chips are created using areas. Overlap operation is utilized to insert the chip areas into the large rectangle.

Main Menu > Preprocessor > Modeling > Create > Areas > Rectangle > By 2 Corners

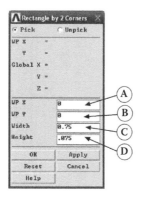

A type 0 in WP X

B type 0 in WP Y

C type 0.75 in Width

D type 0.075 in Height

Apply

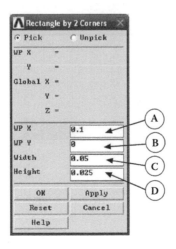

A type 0.1 in WP X

B type 0 in WP Y

C type 0.05 in Width

D type 0.025 in Height

Apply

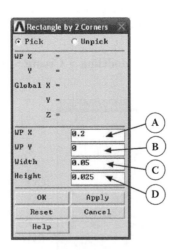

A type 0.2 in WP X

B type 0 in WP Y

C type 0.05 in Width

D type 0.025 in the Height

OK

Main Menu > Preprocessor > Modeling > Operate > Booleans > Overlap > Areas

Pick All

The following step is for changing the material properties of the chips from number 1 to 2. By selecting number 2, the properties of number 2 in the material model are assigned to the chips. The air, by default, has the properties of number 1 in the material model.

Main Menu > Preprocessor > Meshing > Mesh Tool

A select Areas

B click on Set

Using mouse, select both chips

OK

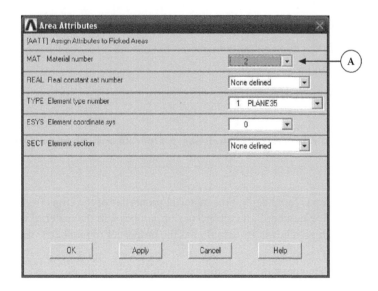

A select 2 in Material number

OK

To ensure that properties of the air and chips are assigned correctly, the air and chips are colored according to their material number in the material model. This step has no effect on the solution.

Utility Menu > PlotCtrls > Numbering...

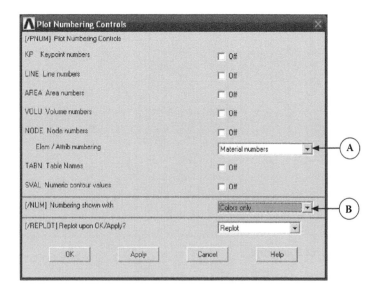

A select Material numbers

B select Colors only

$\boxed{\text{OK}}$

The smart mesh number of 1 has insufficient mesh density to have accurate results for fluid dynamics problems. The elements in the domain can be additionally increased by using the lines size control in the mesh tool. Lines are divided into segments, which will be the number of elements in those lines. The length of the segment is 0.003 m.

Main Menu > Preprocessor > Meshing > Mesh Tool

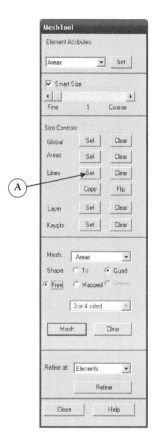

A click the Set in lines

Click on $\boxed{\text{Pick All}}$ to select all lines, and the following window will show up.

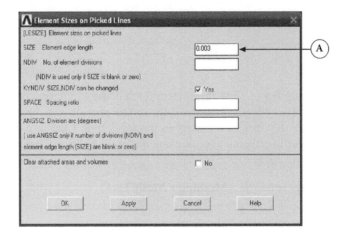

A type 0.003

$\boxed{\text{OK}}$

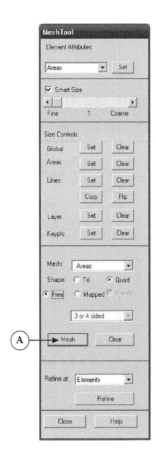

A click on Mesh

Pick All

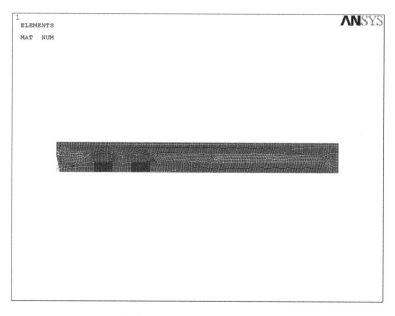

ANSYS graphics shows the mesh

Notice that the ANSYS properties of the air will be used, and the unit is SI. Therefore, the temperature at the boundaries must be in Kelvin scale so that the ANSYS will use the properties at the corresponding temperature. The inlet temperature is 298 K.

Main Menu > Solution > Define Load > Apply > Thermal > Temperature > On Lines

Click at the inlet of the channel

OK

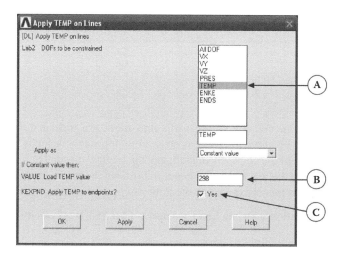

A select TEMP

B type 298 in Load TEMP value

C select Yes

OK

The heat generation must be per unit volume. The applied heat generation is divided by the area of the chip because the problem is two dimensional. The chip's volumetric heat generation is calculated as follows:

$$Q = \frac{10}{0.05 \times 0.025} = 8000 \text{ W/m}^2$$

Main Menu > Solution > Define Load > Apply > Thermal > Heat Generat > On Areas

Click on both chips

OK

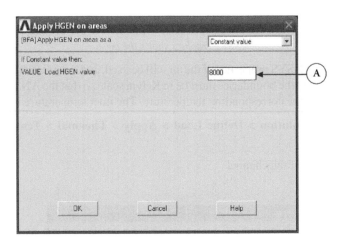

A type 8000 in Load HGEN value

OK

The top and bottom boundaries are walls and these are insulated. Zero x- and y-velocities should be imposed. There is no need to impose the velocity boundary at the bottom surface of the chips. By default, any unassigned thermal boundary condition will be considered as insulated. Do not impose any boundary conditions on the surface of the chips that is exposed to the flow.

Main Menu > Solution > Define Load > Apply > Fluid/CFD > Velocity > On Lines

Click on the top and bottom walls

OK

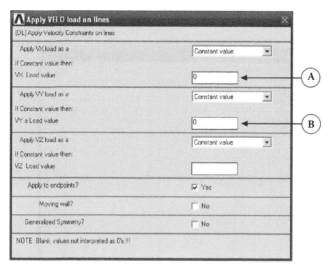

A type 0 in VX Load value

B type 0 in VY Load value

OK

Main Menu > Solution > Define Load > Apply > Fluid/CFD > Velocity > On Lines

Click on the inlet of the channel

OK

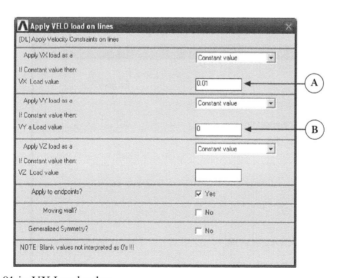

A type 0.01 in VX Load value

B type 0 in VY Load value

OK

Main Menu > Solution > Define Load > Apply > Fluid/CFD > Pressure DOF > On Lines

Click on channel exit

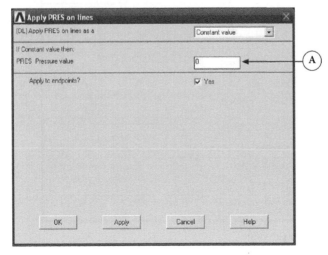

A type 0 in PRES Pressure value

OK

The fluid flow must be solved first, followed by heat transfer. In solution options, the problem will be solved as steady state, and the system is adiabatic, and therefore keeps the default settings. Flow can be changed from laminar to turbulent and from incompressible to compressible in this window.

Main Menu > Solution > FLOTRAN Set Up > Solution Options

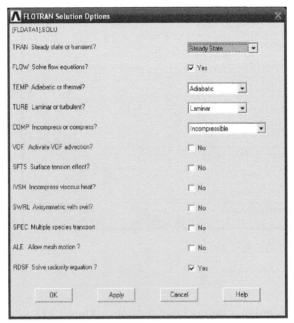

Keep the default settings

OK

In execution control, it is required to specify the number of iterations for fluid flow. Three hundred iterations will be sufficient to reach the convergence criteria for all field variables, which is 1e-6.

Main Menu > Solution > FLOTRAN Set Up > Execution Ctrl

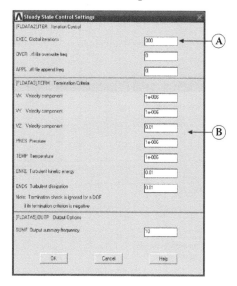

A type 300 in the EXEC Global iteration

B change the Termination Criteria for VX, VY, Pressure, and Temperature to 1e-6

OK

Air properties are assigned in the following step. The SI unit must be selected since the dimensions are in meters, and temperature in Kelvin.

Main Menu > Solution > FLOTRAN Set Up > Fluid Properties

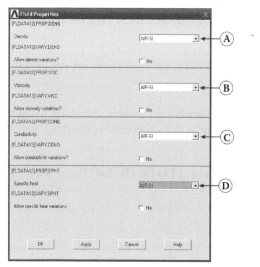

A select AIR-SI in Density

B select AIR-SI in Viscosity

C select AIR-SI in Conductivity

D select AIR-SI in Specific heat

OK

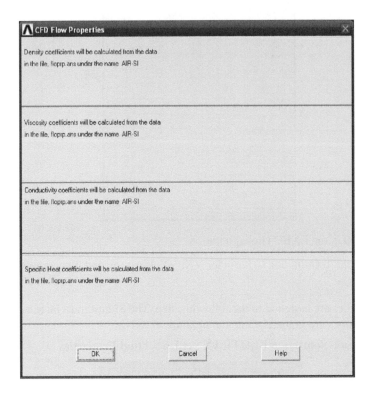

OK

For the heat transfer coefficient calculations, flow bulk temperature is required. In the reference condition, the bulk temperature is specified, which is equal to the flow inlet temperature.

Main Menu > Solution > FLOTRAN Set Up > Flow Environment > Ref Conditions

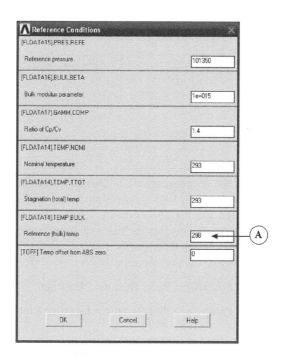

A type 298 in Reference (bulk) temp

OK

Main Menu > Solution > FLOTRAN Set Up > Run FLOTRAN

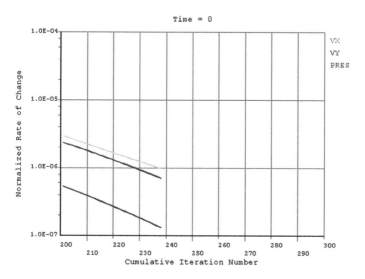

ANSYS graphics shows normalized rate of change for field variables

Close

ANSYS graphics shows the normalized rate of change of the field variables. Note the solution is fully converged. All field variables reach 1e-6 at the iteration number 238, which is less than the specified 300 iterations. Now, the solution is ready for the postprocessor. The velocity contours for the *x*-velocity component is presented to ensure that the fluid flow is solved correctly. In read results, the last set must be called to have the results from the last iteration.

Main Menu > General Postproc > Read Results > Last Set

Main Menu > General Postproc > Plot Results > Contour Plot > Nodal Solu

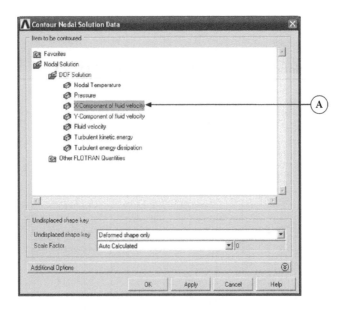

A click on Nodal Solution > DOF solution > X-Component of fluid velocity

OK

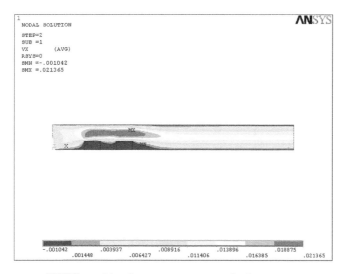

ANSYS graphics shows x-component velocity contours

For the obtained results, the maximum and average velocity in the x-direction occurs above the chips, and the maximum velocity is 0.021365 m/s. This is expected because this region has the smallest cross-section area that forces the flow to accelerate. At the walls, the velocity is zero. The velocity contours are not smooth at the region close to the chips, indicating that more elements are needed, and inaccurate results are expected. Increasing the number of elements can be accomplished using a smaller line division in mesh tools. The fluid part of this problem is completely solved at this stage. Next, the thermal solution will be started. In the solution option, the fluid solver is turned OFF, and the solution is changed to thermal.

Main Menu > Solution > FLOTRAN Set Up > Solution Option

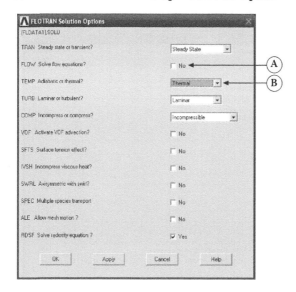

A select No for Solve flow equation

B select Thermal

OK

Main Menu > Solution > FLOTRAN Set Up > Run FLOTRAN

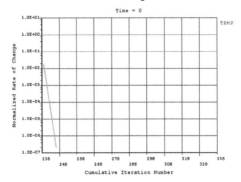

ANSYS graphics shows normalized rate of change for temperature

Close

Main Menu > General Postproc > Read Results > Last Set

Main Menu > General Postproc > Plot Results > Contour Plot > Nodal Solu

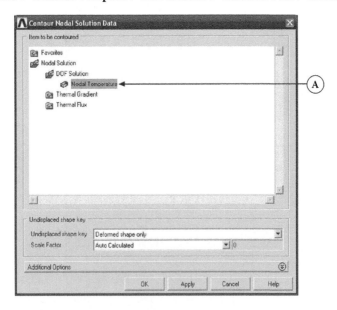

A select Nodal Solution > DOF Solution > Nodal Temperature

OK

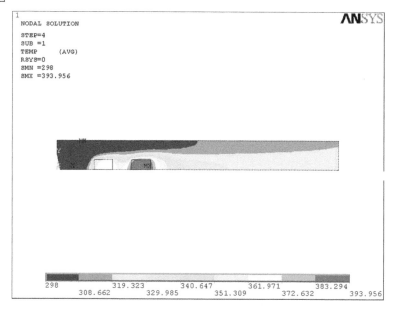

ANSYS graphics shows temperature contours

As shown in the temperature contours, the maximum temperature is 393.956 K, and occurs at the second chip. The path operation is performed in the following steps to obtain the distribution of average heat transfer coefficient around the first chip. The name of the path is arbitrary, and the number of data sets is the maximum number of field variables. The number of divisions is 20 by default, and increasing this number to 50 will produce a more accurate result. A path is created at the surface of the first chip that is exposed to the flow. Then a field variable is assigned into the path. This can be accomplished by using the Map onto Path in the path operation. Only one variable can be selected. Here, the film coefficient is selected.

Main Menu > General Postproc > Path Operation > Define Path > By Location

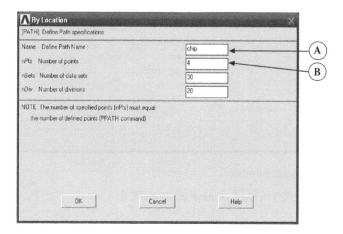

A type chip in the Define Path Name. The name of the path is optional

B type 4 in the Number of points

OK

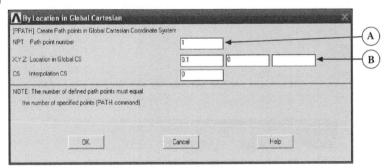

A type 1 in Path point number

B type 0.1 and 0 in the Location in Global CS

OK

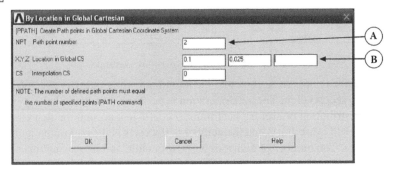

A type 2 in Path point number

B type 0.1 and 0.025 in the Location in Global CS

OK

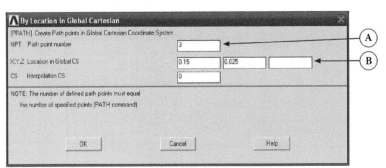

A type 3 in Path point number

B type 0.15 and 0.025 in the Location in Global CS

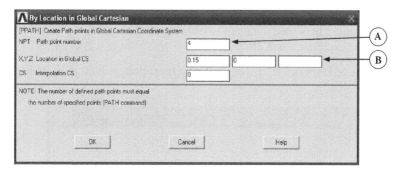

A type 4 in Path point number

B type 0.15 and 0 in the Location in Global CS

OK

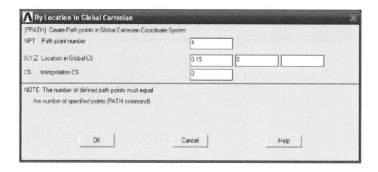

Cancel

Main Menu > General Postproc > Path Operation > Map onto Path

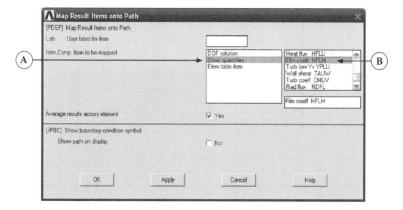

A select Other quantities

B select Film Coeff HFLM

OK

Now, the stored variable, HFLM, is ready to be plotted. In the Plot path item, there are two options. The stored data can be either plotted on graph, or listed. The list results can be exported it into other graphical software such as EXCEL.

Main Menu > General Postproc > Path Operation > Plot path Item > On Graph

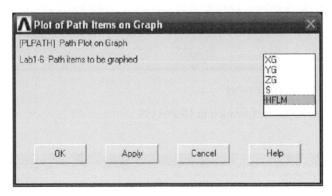

A select HFLM

OK

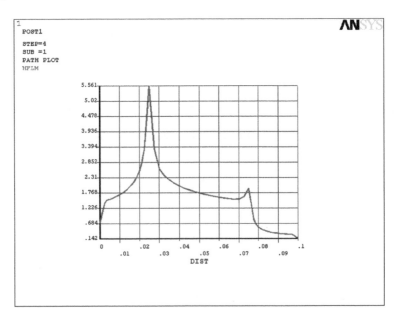

ANSYS graphics shows heat transfer coefficient around the first chip

The average heat transfer coefficient at the chip surface can be determined using the integration in the path operation. The value of the integration must be divided by the path length to get the average value of the variable. Selecting S in Lab2 means that the integration is performed along the path.

Main Menu > General Postproc > Path Operation > Integrate

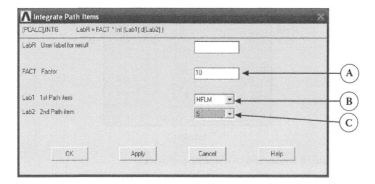

A type 10

B select HFLM

C select S

OK

The ANSYS Output window shows the value of the average heat transfer coefficient around the first chip, which is 1.6569 W/m²-K.

6.4 NATURAL CONVECTION FLOW IN A SQUARE ENCLOSURE

Consider the natural convection flow inside a square enclosure, as shown in Figure 6.6. The left vertical wall is maintained at a constant temperature of 300 K, while the right vertical wall is maintained at 310 K. The bottom and top walls are well insulated. Determine the heat flow at the cold and hot walls. The sides' length of the enclosure is 0.1 m. Use Air properties in the ANSYS.

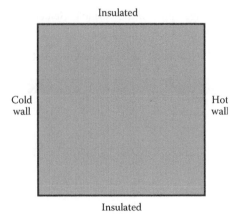

Insulated

Cold
wall

Hot
wall

Insulated

FIGURE 6.6   Natural convection in a square enclosure.

This problem is considered as multiphysics, and physics are coupled, because the fluid flow and heat transfer must be solved at the same time. In this problem, the ANSYS will solve the heat transfer first, and then it uses the temperature to determine the fluid properties for the fluid flow. The fluid flow solution will be used to solve for heat transfer again. This process is continued until the solution for both physics is fully converged.

Double click on the ANSYS icon

Main Menu > Preferences

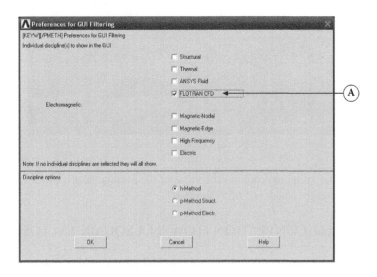

A select the FLOTRAN CFD

Main Menu > Preprocessor > Element type > Add/Edit/Delete

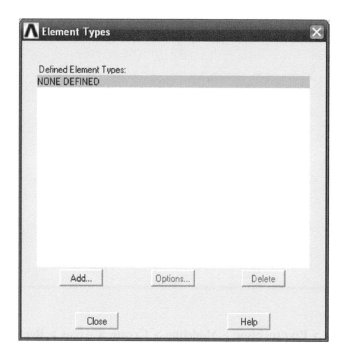

Add...

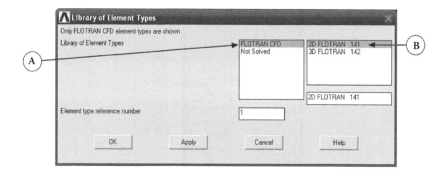

A select FLOTRAN CFD

B select 2D FLOTRAN 141

OK

Close

Main Menu > Preprocessor > Modeling > Create > Areas > Rectangle > By 2 Corners

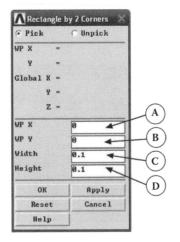

A type 0 in WP X

B type 0 in WP Y

C type 0.1 in Width

D type 0.1 in the Height

OK

Coupled physics, such as natural convection flow, requires the mesh to be fine. The smart mesh of number 1 still provides insufficient number of elements in the domain. The number of elements can be increased if a manual sizing is used and dividing the lines to small lines, such as 100. For this technique, do not turn the smart mesh ON in the mesh tool.

Main Menu > Preprocessor > Meshing > Size Cntrls > ManualSize > Lines > All Lines

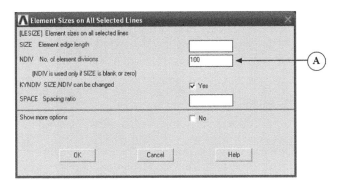

A type 100 in No. of element divisions

OK

Main Menu > Preprocessor > Meshing > Mesh Tool

A Mesh

Pick All

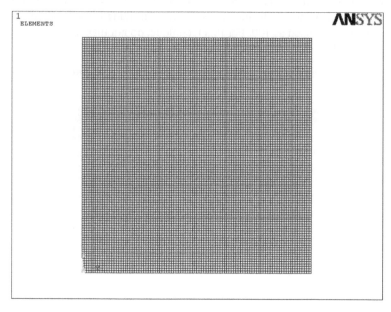

ANSYS graphics shows the mesh

Main Menu > Solution > Define Load > Apply > Thermal > Temperature > On Lines

Click on left line

OK

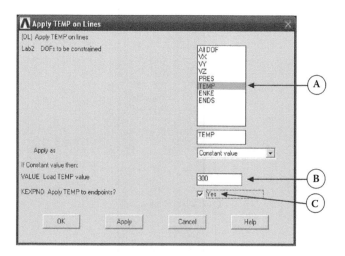

A select TEMP

B type 300 in Load TEMP value

C select Yes

OK

Main Menu > Solution > Define Load > Apply > Thermal > Temperature > On Lines

Click on right line

OK

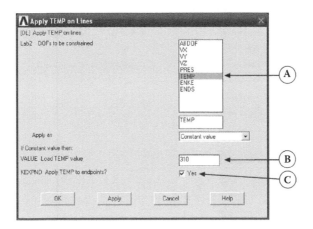

A select TEMP

B type 310 in Load TEMP value

C select Yes

OK

Main Menu > Solution > Define Load > Apply > Fluid/CFD > Velocity > On Lines

Click on the left, right, top, and bottom walls

OK

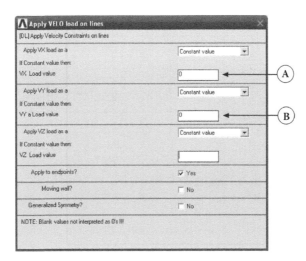

A type 0 in VX Load value

B type 0 in VY Load value

$\boxed{\text{OK}}$

Main Menu > Solution > FLOTRAN Set Up > Solution Options

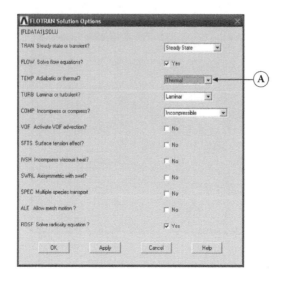

A Select Thermal

$\boxed{\text{OK}}$

Main Menu > Solution > FLOTRAN Set Up > Execution Ctrl

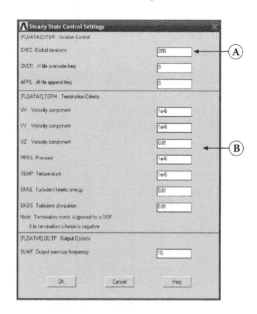

A type 200 in the EXEC Global iteration

B change the Termination Criteria for VX, VY, Pressure, and Temperature to 1e-6

OK

The natural convection flow inside the enclosure is due to the variation of density with temperature. Hence, making density the function of temperature is essential to generate the natural convection flow. Additionally, the acceleration gravity in the y-direction is also important to generate the natural convection flow. If density variation is not turned ON or gravity acceleration is not assigned, the problem will be solved as only a pure conduction problem.

Main Menu > Solution > FLOTRAN Set Up > Fluid Properties

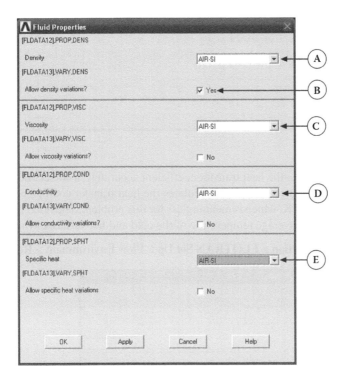

A select AIR-SI in Density

B click on Yes for Allow density variation

C select AIR-SI in Viscosity

D select AIR-SI in Conductivity

E select AIR-SI in Specific heat

OK

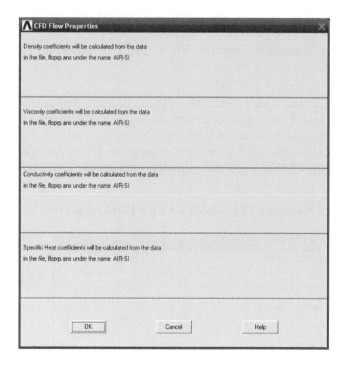

OK

In order to calculate the heat transfer coefficient accurately, a correct fluid reference temperature is required. ANSYS calculates the heat transfer coefficient based on a temperature of 298 K, which is meaningless for this problem. The reference temperature should be the average temperature of the cold and hot surfaces, which is 305 K.

Main Menu > Solution > FLOTRAN Set Up > Flow Environment > Ref Condition

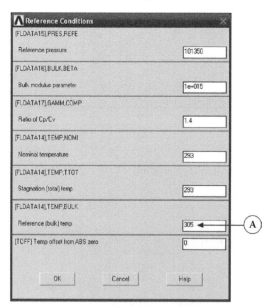

A type 305 in the Reference (bulk) temp

OK

Main Menu > Solution > FLOTRAN Set Up > Flow Environment > Gravity

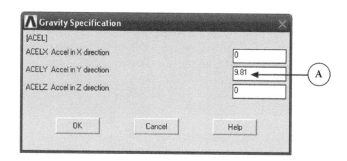

A type 9.81 in ACELY Accel in Y direction

OK

Main Menu > Solution > FLOTRAN Set Up > Run FLOTRAN

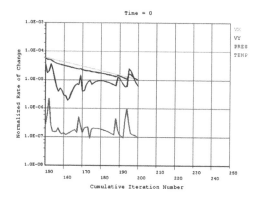

ANSYS graphics shows normalized rate of change for field variables

Close

The Normalized rate of change indicates that the solution did not converge. The velocity components and pressure reach 10^{-5}, and therefore additional iterations are required.

Main Menu > Solution > FLOTRAN Set Up > Run FLOTRAN

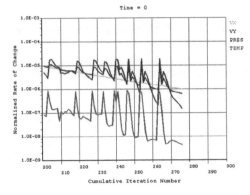

ANSYS graphics shows normalized rate of change for field variables

Close

Now, the solution is fully converged. The next step is to inspect the temperature contours and velocity vectors to check if the problem is correctly solved.

Main Menu > General Postproc > Read Results > Last Set

Main Menu > General Postproc > Plot Results > Contour Plot > Nodal Solu

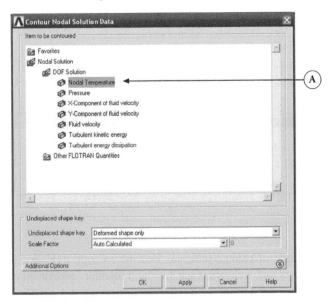

A click on Nodal Solution > DOF Solution > Nodal Temperature
OK

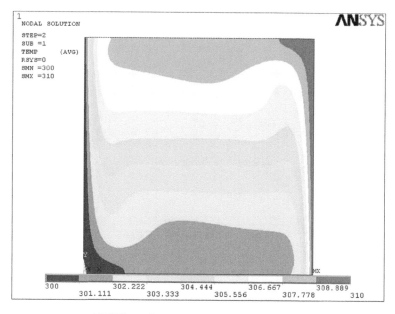

ANSYS graphics shows temperature contours

Main Menu > General Postproc > Plot Results > Vector Plot > Predefined

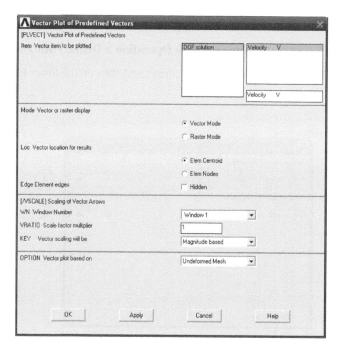

OK

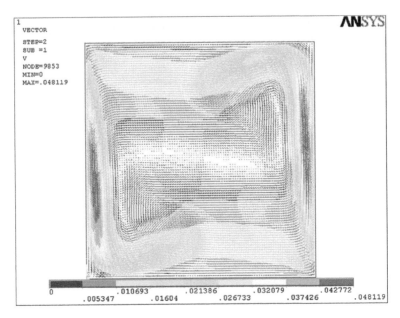

ANSYS graphics shows velocity vectors

From the temperature contours, left wall is kept at a low temperature, while the right wall is kept at high temperature. The velocity vector figure shows the air flowing downward at the cold wall region, while the air flowing upward at the hot wall region. The region at the center of the enclosure is nearly stagnant. The problem is solved correctly. The following steps are for fining the average heat flow at the cold and hot walls.

Main Menu > General Postproc > Path Operation > Define Path > by nodes

Using the mouse, click on the upper left corner, and then on the lower left corner.

OK

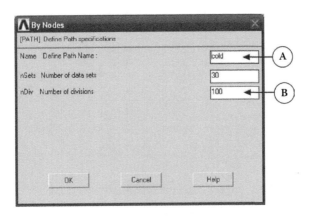

A name the path as Cold

B type 100 in nDiv Number of divisions
$\boxed{\text{OK}}$

The name of the path is arbitrary. The number of data sets is the maximum number of field variables. The number of divisions is 20 by default, and increasing this number to 100 will produce result that is more accurate. Next step is assigning a field variable into the path. This can be accomplished by using the Map onto Path in the path operation. Only one variable can be selected at a time. Here, the heat transfer coefficient is selected.

Main Menu > General Postproc > Path Operation > Map onto Path

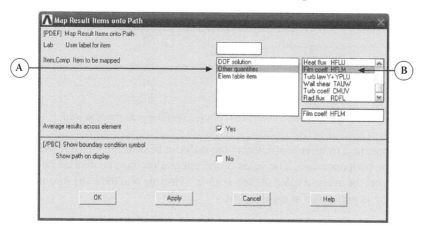

A select Other quantities

B select Film Coeff HFLM
$\boxed{\text{OK}}$

Main Menu > General Postproc > Path Operation > Plot path Item > On Graph

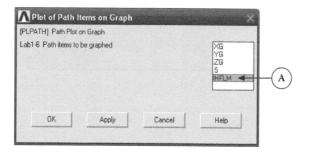

A select HFLM
$\boxed{\text{OK}}$

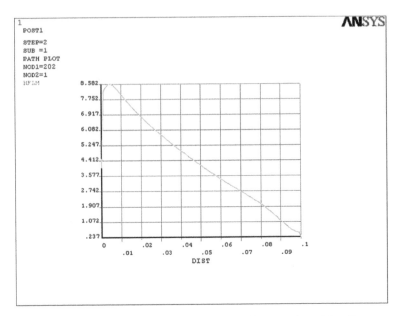

ANSYS graphics shows heat transfer coefficient at the cold wall

The average heat transfer coefficient at the wall can be determined using the integration in the path operation. The value of the integration must be divided by the path length to get the average value of the variable. Selecting S in the Lab2 means that the integration is performed along the path.

Main Menu > General Postproc > Path Operation > Integrate

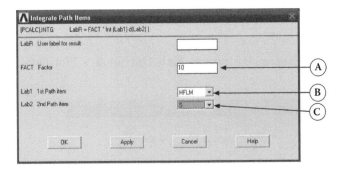

A type 10

B select HFLM

C select S

OK

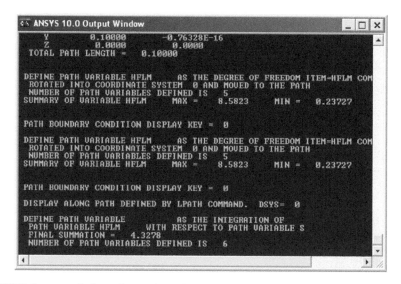

ANSYS Output window shows the value of the average heat transfer coefficient at the cold wall which is 4.3278 W/m²-K. Hence, the heat flows out of the cold wall is

$$Q_C = hA(T_c - T_m) = 4.3278 \times 0.1 \times (300 - 305) = -2.1639 \text{ W}$$

Now the average heat transfer coefficient at the hot wall is calculated in the following steps.

Main Menu > General Postproc > Path Operation > Define Path > by nodes

Using the mouse, click on the upper right corner, and then on the lower right corner.

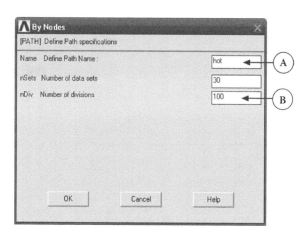

A name the path as hot

B type 100 in nDiv Number of divisions

Main Menu > General Postproc > Path Operation > Map onto Path

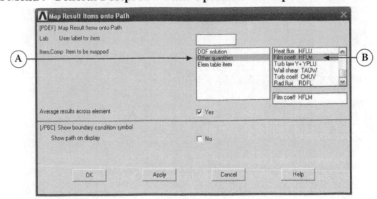

A select Other quantities

B select HFLM

OK

Main Menu > General Postproc > Path Operation > Plot path Item > On Graph

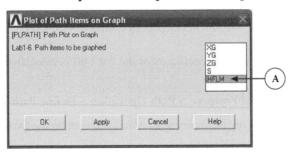

A select HFLM

OK

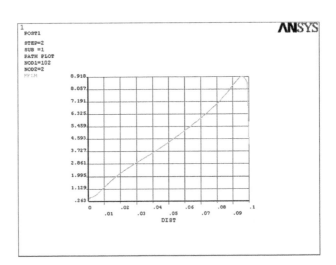

ANSYS graphics shows heat transfer coefficient at the hot wall

Main Menu > General Postproc > Path Operation > Integrate

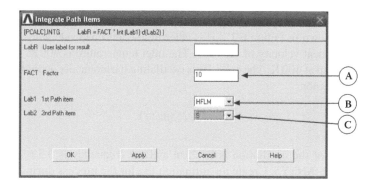

A type 10

B select HFLM

C select S

OK

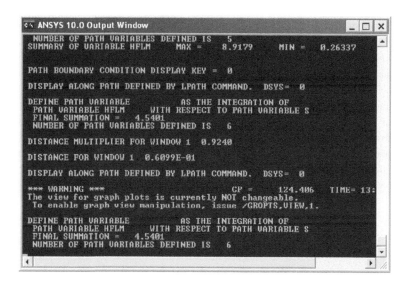

ANSYS Output window shows the value of the average heat transfer coefficient at the hot wall which is 4.5401 W/m²-K. Hence, the heat flow out of the hot wall is

$$Q_H = hA(T_H - T_m) = 4.5401 \times 0.1 \times (310 - 305) = 2.2701 \text{ W}$$

Heat flow into the cold wall, Q_H, and heat flow out of the hot wall, Q_C, are almost the same. The observed small error is due to numerical calculations, and this error can be minimized if a finer mesh is used.

6.5 OSCILLATIONS OF A SQUARE-HEATED CYLINDER IN A CHANNEL

Consider an oscillating square cylinder with a cross section of 0.025 m × 0.025 m in a channel, as shown in Figure 6.7. Air is entering the left open side with a uniform and time-dependent velocity of 0.05 m/s. The inlet temperature is 300 K, while the cylinder is heated at 400 K. The cylinder is oscillating harmonically according to the following expression:

$$y(t) = 0.025 \sin(\pi t)$$

The amplitude of the oscillation is 0.025 m and the frequency is 0.5 s⁻¹. Use Air properties in the ANSYS. Create an animation file for the process during the time interval of 0–5 s.

    The forced convection flow in a channel containing bluff objects has received increased attention in the last decade because of their importance in engineering applications. The geometry of the horizontal channel with bluff is commonly found in heat exchanger devices. On the other hand, the flow structures are not completely understood, especially of the unsteady flow. Most of the experimental or numerical investigations of forced convection flow in pertain to flow in fixed bluff objects. The oscillating bluff object enhances the heat flow in the channel by mixing and reducing the temperature gradients in the domain. The example focuses on the flow and thermal characteristics of the square cylinder.

    The physical model used in this work is shown in Figure 6.7. A two-dimensional horizontal channel with a square cylinder is used. The outer walls of the channel are fixed, while the square cylinder is oscillating vertically with a frequency of 0.5 s⁻¹ and a maximum amplitude of 0.025 m. The working fluid is air. Initially, the square cylinder is at the center of the channel and the air's temperature is equal to the inlet temperature. To simulate the motion of the inner cylinder, the ALE kinematics description method is utilized. In this method, the computational mesh can be moved with fluid (Lagrangian), fixed (Eulerian), or moved in a prescribed way. Therefore,

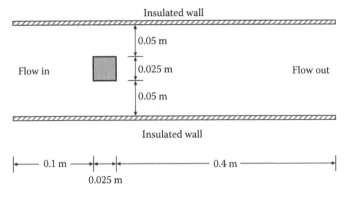

FIGURE 6.7   Oscillating heated square cylinder in a channel.

the ALE method is used to numerically investigate the effect of oscillations of the inner cylinder upon the flow field and heat transfer characteristics. The Eulerian equations of motion need to be modified to reflect the moving frame of reference. The time derivative terms need to be written in the term of the moving frame of reference, as follows:

$$\frac{\partial \Pi}{\partial t}\bigg|_{\text{fixed frame}} = \frac{\partial \Pi}{\partial t}\bigg|_{\text{moving frame}} - \vec{W} \cdot \nabla \Pi$$

where
  Π is any degree of freedom
  $\vec{W}$ is the velocity of the moving frame of reference

At the steady state, the Eulerian sense requires

$$\frac{\partial \Pi}{\partial t}\bigg|_{\text{fixed frame}} = 0$$

This problem is considered as multiphysics and physics are coupled because the fluid flow and heat transfer must be solved at the same time. In this problem, the ANSYS will solve the fluid flow and heat transfer first, and then the mesh is moved according to the user's function. The fluid flow and heat transfer will be solved again for the new mesh.

Double click on the ANSYS icon

Main Menu > Preferences

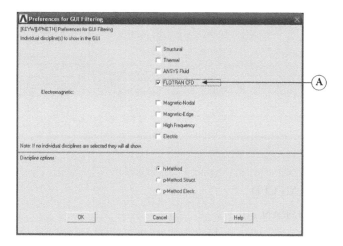

A select the FLOTRAN CFD

Main Menu > Preprocessor > Element type > Add/Edit/Delete

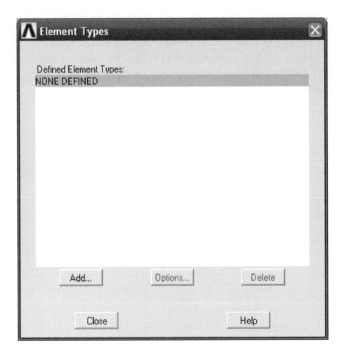

Add...

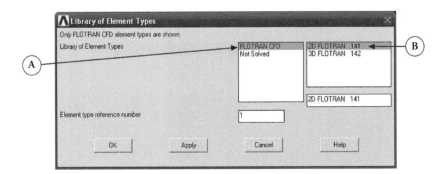

A select FLOTRAN CFD

B select 2D FLOTRAN 141

OK

This problem involves a mesh movement. In the options, the support mesh displacement should be enabled.

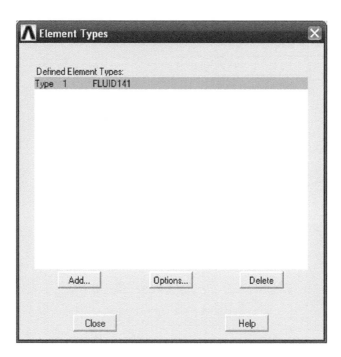

Option

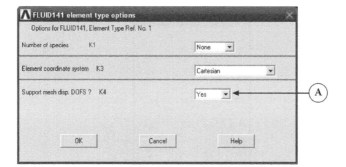

A Select Yes in Support mesh disp.

OK

Close

Main Menu > Preprocessor > Modeling > Create > Areas > Rectangle > By 2 Corners

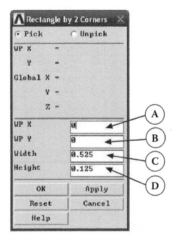

A type 0 in WP X

B type 0 in WP Y

C type 0.525 in Width

D type 0.125 in the Height

| Apply |

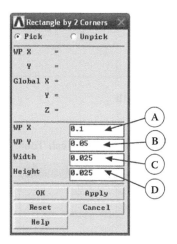

A type 0.1 in WP X

B type 0.05 in WP Y

C type 0.025 in Width

D type 0.025 in the Height

| OK |

Main Menu > Preprocessor > Modeling > Operate > Booleans > Overlap > Areas

click on | Pick All | to select all areas

Main Menu > Preprocessor > Modeling > Delete > Area Only

Click on the square to highlight it

| OK |

Coupled physics, such as moving mesh problems, requires the mesh to be fine. The smart mesh of number 1 still provides only an insufficient number of elements in the domain. The number of elements can be increased if a manual sizing is used and dividing the lines to small segments. For this technique, do not turn the smart mesh ON in the mesh tool. All lines are divided into segments with 0.005 m length each. A smaller line segment would be better, but it will take much longer computational time.

Main Menu > Preprocessor > Meshing > Size Cntrls > ManualSize > Lines > All Lines

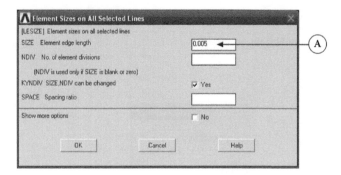

A type 0.005 in Element edge length

OK

Main Menu > Preprocessor > Meshing > Mesh Tool

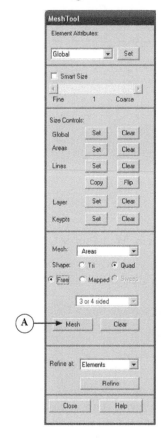

A Mesh

Pick All

Close

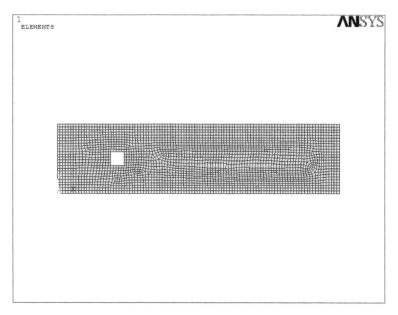

ANSYS graphics shows the mesh

Main Menu > Solution > Define Load > Apply > Thermal > Temperature > On Lines

Click on inlet of the channel

OK

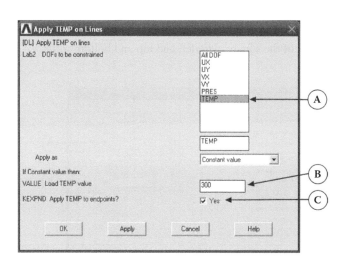

A select TEMP

B type 300 in Load TEMP value

C select **Yes**

OK

Main Menu > Solution > Define Load > Apply > Thermal > Temperature > On Lines

Click on the lines of the square

OK

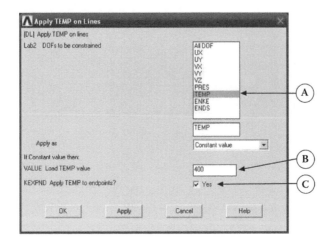

A select TEMP

B type 400 in Load TEMP value

C select Yes

OK

Main Menu > Solution > Define Load > Apply > Fluid/CFD > Velocity > On Lines

Click on the walls of the square cylinder, and top and bottom walls

OK

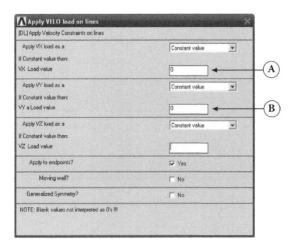

A type 0 in VX Load value

B type 0 in VY Load value

OK

Main Menu > Solution > Define Load > Apply > Fluid/CFD > Velocity > On Lines

Click on the inlet of the channel

OK

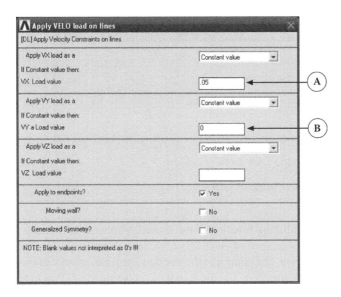

A type 0.05 in VX Load value

B type 0 in VY Load value

OK

Main Menu > Solution > Define Load > Apply > Fluid/CFD > Pressure > On Lines

Click on the exit of the channel

OK

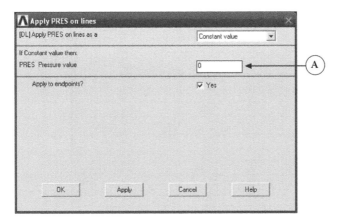

A type 0 in Pressure value

OK

The initial condition for the velocity components, displacements, and pressure is zero, while the initial condition for temperature is 300 K. By default, the initial condition for all field variables is zero. Hence, only assign the initial condition for temperature.

Main Menu > Solution > Define Load > Apply > Initial Condition > Define

Pick All

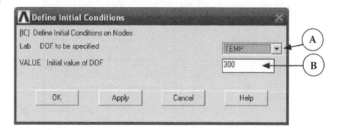

A select TEMP in the DOF to be specified

B type 300 in Initial value of DOF

OK

The sinusoidal oscillating function of the square cylinder can be easily applied with the function editor feature of ANSYS. ANSYS will create a data table from the displacement function, and apply it to the selected lines. The mesh will move according to the displacement function. To ensure that the equation is written correctly, a graph of the displacement is created in the ANSYS graphics.

Main Menu > Solution > Define Load > Apply > Functions > Define/Edit

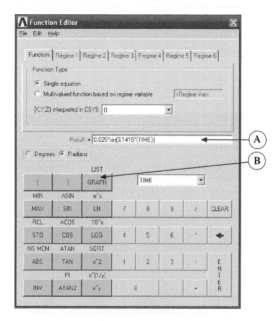

A Type the equation of the oscillating: 0.025*sin(3.1415*{TIME})

B click on GRAPH

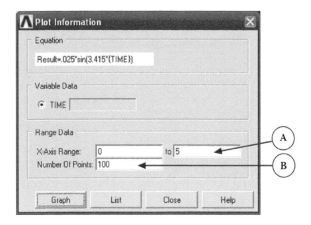

A type 0 and 5 for X-Axis Range

B type 100 in the Number Of Points

Graph

0 and 5 are the range of the data in the *x*-axis, while 100 is the number of data to be plotted. Number Of Points has nothing to do with the accuracy of the results, and Higher number of it will just create a smother plot.

Close

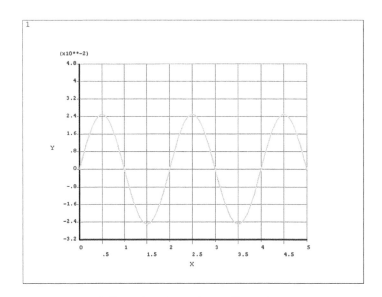

ANSYS graphics shows the oscillating function

In the Equation Editor click on File then Save

Save the file as DisY, the file name is optional

Save

After saving the function, it is required to load it to the ANSYS solution using the read file comments.

Close the Function Editor window

Main Menu > Solution > Define Load > Apply > Functions > Read File

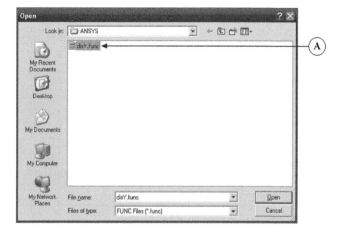

A select disY.func

Open

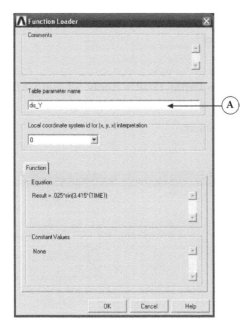

A type dis_Y in the Table parameter name

OK

The dis_Y will be shown later when applying the boundary conditions in the existing table. The lower and upper walls are fixed, while the square cylinder walls are displaced according to the input equation.

Main Menu > Solution > Define Load > Apply > Fluid/CFD > Displacement > On Lines

Click on the upper and lower walls

OK

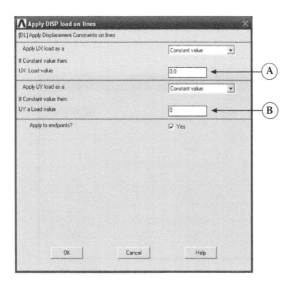

A type 0 in UX Load value

B type 0 in UY Load value

$\boxed{\text{OK}}$

Main Menu > Solution > Define Load > Apply > Fluid/CFD > Displacement > On Lines

Click on the lines of the square

$\boxed{\text{OK}}$

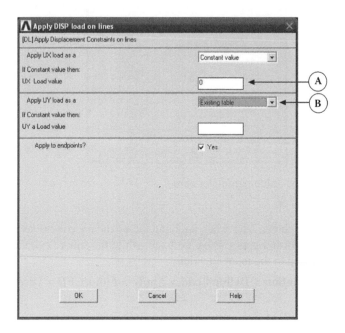

A type 0 in UX Load value

B select Existing table

$\boxed{\text{OK}}$

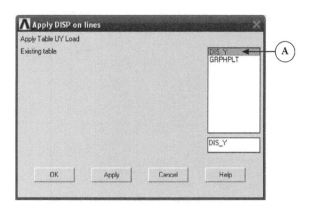

A type DIS_Y

$\boxed{\text{OK}}$

The following step is very important to initialize the moving mesh. The problem must be transient, and flow and thermal analysis must be enabled. ALE should be turned ON to allow mesh motion. If the ALE is not checked, the square cylinder will not move.

Main Menu > Solution > FLOTRAN Set Up > Solution Options

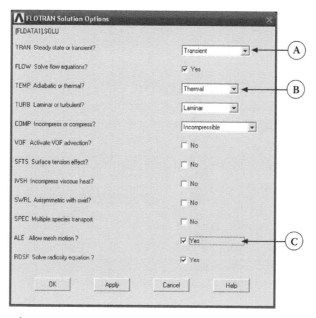

A Select Transient

B Select Thermal

C click on ALE mesh motion

$\boxed{\text{OK}}$

The total time for the problem is 5 s, and selected time step size is 0.1 s. The small time step size is used to ensure the convergence at each time step. The time interval of 0.1 s is used to store the results at each time step.

Main Menu > Solution > FLOTRAN Set Up > Execution Ctrl

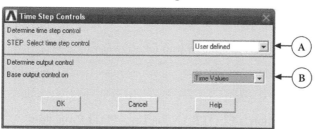

A select User defined

B select Time values

OK

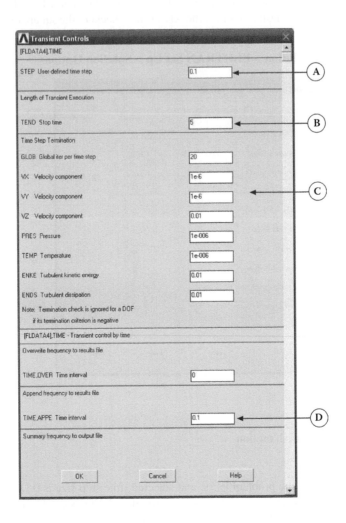

A type 0.1 in User-defined time step

B type 5 in Stop time

C set the Time Step Termination to 1e-6 for all field variables

D Type 0.1 in Time interval

OK

Main Menu > Solution > FLOTRAN Set Up > Fluid Properties

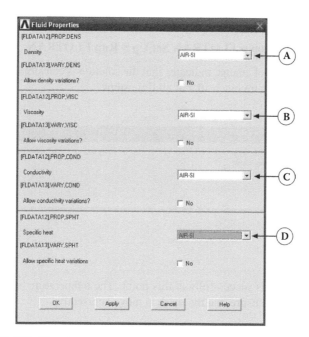

A select AIR-SI in Density

B select AIR-SI in Viscosity

C select AIR-SI in Conductivity

D select AIR-SI in Specific heat

OK

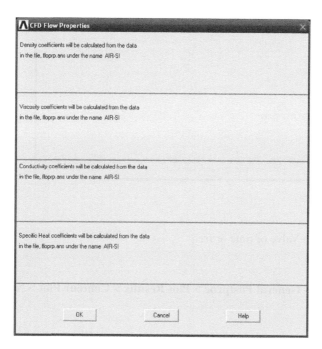

OK

Main Menu > Solution > FLOTRAN Set Up > Run FLOTRAN

The Normalized rate of change indicates that the solution is converged. The velocity components and pressure reach 10^{-6} at all time steps.

Close

The solution task ends successfully at this point. The temperature contours at time 4.95 s is presented to ensure that the mesh is moved correctly.

Main Menu > General Postproc > Read Results > by Time/Freq

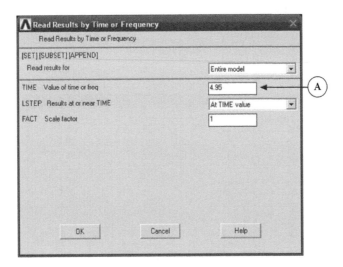

A type 4.95 in Valve of time or freq

OK

Main Menu > General Postproc > Plot Results > Contour Plot > Nodal Solu

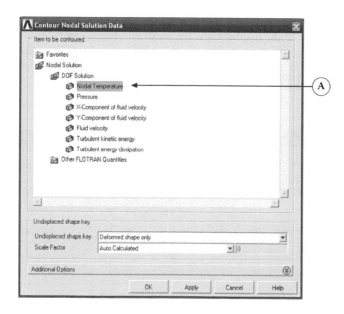

A click on Nodel Solution > DOF Solution > Nodal Temperature

OK

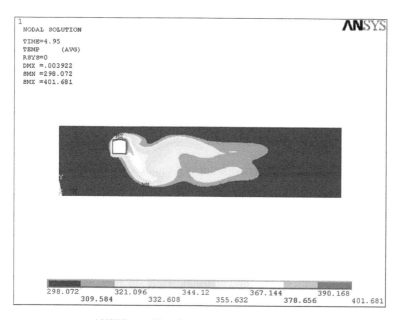

ANSYS graphics shows temperature contours

Notice that the square cylinder is moved upward. The following steps are for creating the animation file. In the display type, the mesh movement can be animated as well as all field variables. Here, the animation for temperature contours is selected.

Main Menu > General Postproc Utility Menu > PlotCtrls > Animate > Over time …

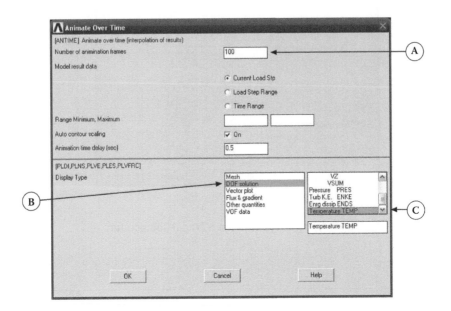

A type 100 in Number of animation frames

B select DOF solution

C select Temperature

$\boxed{\text{OK}}$

ANSYS will create an animation file for the square oscillation with temperature contours.

PROBLEMS

6.1 A pipeline used for transporting highly corrosive liquids is made of concentric carbon steel and aluminum pipes, as shown in Figure 6.8. The inner pipe is made of aluminum to prevent corrosion, while the outer pipe is made of carbon steel to add strength to the pipeline. At the inner surface of the pipe, the temperature is 125°C, while at the outer surface the temperature is 20°C. As a result, a thermal stress is induced at the contact surface that could damage the pipeline. Determine the maximum stress in the pipe. Mesh the computational domain with a smart mesh size of 1.

Property	Aluminum	Carbon Steel
Thermal conductivity (W/m-°C)	83	111
Young modulus (Pa)	70×10^9	210×10^9
Poisson's ratio	0.33	0.29
Thermal expansion (1/K)	23×10^{-6}	12×10^{-6}

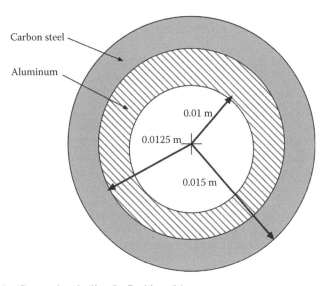

FIGURE 6.8 Composite pipeline for Problem 6.1.

6.2 For an electronic chip mounted on a channel shown in Figure 6.9, the working fluid is air, and the thermophysical properties are listed below. The inlet velocity is 0.01 m/s, while the exit velocity is the atmospheric pressure. The inlet temperature is 20°C, and 3 W is generated only in the heat source, as shown in the figure. Mesh the computational domain with a smart mesh size of 1.

Property	Chip	Air
Density (kg/m³)	850	1.21
Conductivity (W/m-°C)	1.5	0.025
Specific heat (J/kg-°C)	900	1010
Viscosity (Pa-s)	–	18.26e-6

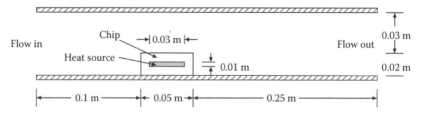

FIGURE 6.9 A chip in a channel for Problem 6.2.

Determine

1. Maximum temperature in the domain
2. Maximum heat transfer coefficient around the chip surface
3. Average heat transfer coefficient around the chip
4. Show that the energy balance is satisfied

6.3 Air is heated in a channel using four circular heaters, as shown in Figure 6.10. Heat is volumetrically generated in the heaters, 5 W in each heater. Flow velocity at the inlet of the channel is 0.05 m/s and temperature is 300 K. Determine the exit temperature of the air. The properties of the heaters are $\rho = 950\,\text{kg/m}^3$, $C_p = 1250\,\text{J/kg-K}$, $k = 35\,\text{W/m-K}$.

6.4 Consider the natural convection flow inside a concentric annulus enclosure, as shown in Figure 6.11. The inner cylinder is maintained at a constant temperature of 310 K, while the outer cylinder is maintained at 315 K. The working fluid is air, and use air properties in the ANSYS. Determine the heat flow at the inner and outer cylinders.

6.5 Consider the two circular cylinders with diameters of 0.025 m × 0.025 m, as shown in Figure 6.12. Air is entering the left open side with a uniform velocity of 0.1 m/s. The inlet temperature is 300 K, while the cylinder's wall is heated at 350 K. The initial temperature of the air is 300 K. The first cylinder is oscillating harmonically according to the following expression:

$$y(t) = 0.05\sin(2\pi t)$$

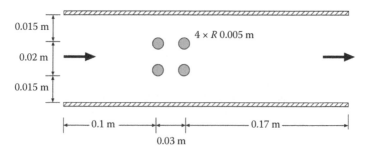

FIGURE 6.10 A channel with four circular heaters for Problem 6.3.

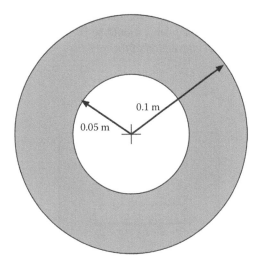

FIGURE 6.11 Natural convection in an annulus enclosure for Problem 6.4.

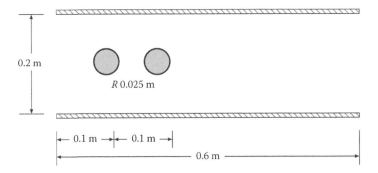

FIGURE 6.12 An oscillating cylinder in cross flow for Problem 6.5.

Create an animation file for the process during the interval of 0–7 s. Also, calculate the flow exit temperature at time = 3, 5, and 7 s.

6.6 Figure 6.13 shows a square enclosure containing air, and its external wall is subjected to a free convection of 300 K, and 5 W/m-K. Inside, a rectangular object is maintained at a uniform temperature of 310 K, and oscillating harmonically according to the relationship:

$$y(t) = 0.05\sin(\pi t)$$

The initial temperature of the air is 300 K. Create an animation file for the process during the interval of 0–5 s. The results should be presented for two cases: with and without the effect of natural convection of the air.

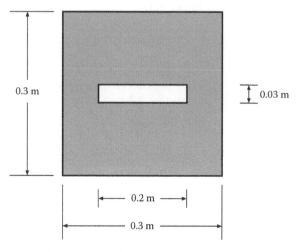

FIGURE 6.13   An oscillating rectangular object inside a square enclosure for Problem 6.6.

7 Meshing Guide

7.1 MESH REFINEMENT

The accuracy of the finite element model can be enhanced by either increasing the number of elements in the model or using higher-order elements. Increasing the number of element is called h-refinement, while increasing the order of the element is called p-refinement. Increasing the number of elements or the elements' order will lead to significant increase in the computational time and requires memory to solve the problem. Consider the stress analysis of a plate with holes shown in Figure 7.1a. A linear quadratic element is used to determine the stress concentration factor. The exact solution is available from the theory of elasticity, and Figure 7.1b shows the finite element mesh. The mesh consists of 86 elements. To study the effect of the number of elements on the solution, the number of elements is increased and the stress concentration is obtained for all mesh sizes.

Figure 7.2 shows the error of the stress concentration for different number of elements using linear quadratic elements. The domain is meshed with three different mesh sizes, A, B, and C. The number of elements of mesh C is higher than B, and B is higher than A. The figure indicates that as the number of elements increases, the relative error is decreased to approach a fixed value. Adding more elements to the mesh C will have an insignificant effect in reducing the computational error.

When the number of elements has no effect on the solution, the mesh is called a mesh-independent solution. Increasing the order of the elements reduces error percentage for the same number of elements. Figure 7.3 indicates that replacing the linear quadratic elements with a cubical one reduces the percentage of the error by about 10%.

With a high-order element, the mesh-independent process is faster than a low-order element, as shown in Figure 7.4. In addition, a mesh with the linear quadratic element is becoming mesh independent at a very slow rate.

The higher-order elements lead to more accurate results than lower-order elements. On the other hand, the time required to complete the solution is important, especially for very large meshes. The computational time could take more than 10 days for meshes with more than one million elements. In addition to time, the higher-order elements need large amount of memory that could not be afforded. Figure 7.5 indicates that in spite of a low mesh convergence rate of linear quadratic elements, the time required to solve the problem is nearly independent of the number of elements in the mesh. Also, increasing the order of the elements significantly increases the time required to solve the problem.

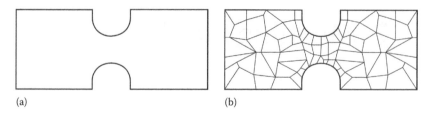

(a)   (b)

FIGURE 7.1   (a) Geometry of a plate, and (b) finite element mesh.

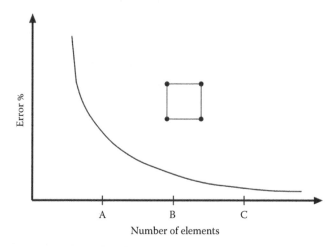

FIGURE 7.2   Error of the stress concentration for different mesh size using linear quadratic elements.

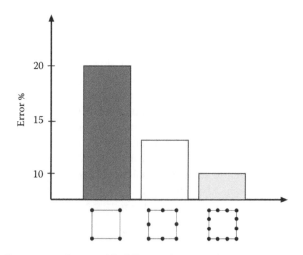

FIGURE 7.3   Percentage of error with different element orders.

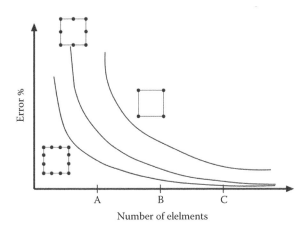

FIGURE 7.4   Error of the stress concentration for different number of elements with different orders.

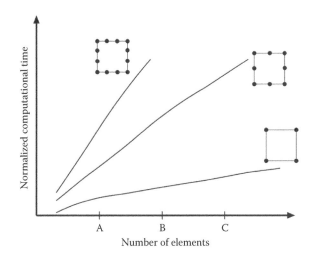

FIGURE 7.5   Computational time for various types and numbers of quadrilateral elements.

7.2   ELEMENT DISTORTION

When a complex geometry is meshed using an automatic mesh generator, elements at some critical regions in the geometry become distorted, and the accuracy of the results is significantly decreased at these regions. Hence, this section concerns with the element distortions, and discusses finite element meshes and identifies the bad elements. In addition, guidelines on good element shapes are also introduced in this section. There are four types of element distortion:

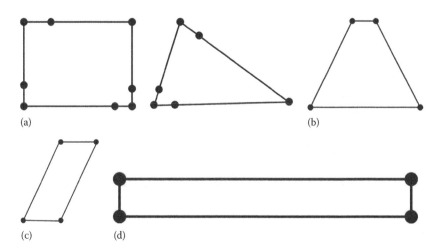

(a)   (b)

(c)   (d)

FIGURE 7.6 The major types of element distortion: (a) mid-side node off-center, (b) taper, (c) skew, and (d) aspect ratio.

1. Mid-side node off-center
2. Taper
3. Skew
4. Aspect ratio

The mid-side node off-center type is when side nodes are shifted from the center of the side. As the node moves away from the center, the errors increase. In practice, the mid-side node should not be shifted more than one-eighth of the side's length. Rectangular elements can also be skewed or tapered. The angles of the rectangular elements should be 90°. For a rectangular element, when the height to width ratio exceeds 2, the element has an aspect ratio problem. It turns out that a perfectly shaped equilateral triangle should have an aspect ratio of 1.15, and a square element should have an aspect ratio of 1. Aspect ratio higher than the recommended one can only occur in noncritical regions in the model, such as in uniform temperature or stress regions. Figure 7.6 shows the major types of element distortion. Element distortions are not limited to two-dimensional spaces, they can also occur in three-dimensional spaces. The most common one is the warping. Warping occurs when the faces of a solid element do not lie in the same plane. The warping angle should not be more than 10°; otherwise an error is expected.

7.3 MAPPED MESH

Meshes can be classified as a free or a mapped mesh. The elements of the free mesh are randomly distributed in the domain with different sizes and shapes, and Figure 7.7 shows a free mesh for a wavy channel. Element distortions are most likely to appear in these meshes. Most finite element software generates these meshes. Poor results are always associated with free meshes. The free meshes should be only limited to very complex geometries.

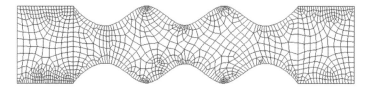

FIGURE 7.7   Free mesh for a wavy channel.

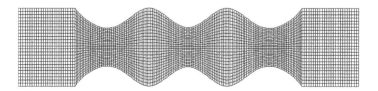

FIGURE 7.8   Mapped mesh of a wavy channel.

On the other hand, mapped meshes have organized elements and element sizes are gradually changing in the domain, and the mesh is free of the element distortions. Figure 7.8 shows a mapped mesh of a wavy channel. The user can fully control the size of the elements and their distribution in the computational domain. These meshes are typically used in finite element simulations because of their flexibility and ability to mesh complex geometries. Mesh designers should have a strong understanding of the physics of the problem before generating the mapped mesh. For example, for an elasticity problem, the areas with high and uniform stress gradients should be located, and design the mesh accordingly at these areas.

For solid mechanics problems, elements should be concentrated at the regions with stress gradients. For example, for a plate with a hole, as shown in Figure 7.9, stress is extremely high at regions close to the hole. Therefore, the mesh is dense at these regions. For fluid problems, the regions close to the walls have a very high velocity gradient due to shear forces, and the elements should be dense at regions close to the walls.

Figure 7.10 shows a mapped mesh of channel containing multiple blocks. Elements are concentrated near the region close to the blocks' surface to capture some critical features, such as the heat transfer coefficient and shear stress.

The geometry shown in Figure 7.11 is for a fin, and a thermal analysis is required. High temperature gradients at the upper region are expected. Therefore, elements must be concentrated at this region. The geometry is meshed with three different mapped meshes. The mesh in Figure 7.11a is unacceptable because the elements are not concentrated in the upper region, and poor results are expected. In the second mesh, Figure 7.11b, the upper region of the geometry has more elements than the first mesh, but an abrupt change in element size should be avoided, because any errors in the large elements will transfer into the small elements. In addition, matrix operations of these meshes are difficult. The third mesh, Figure 7.11c, is the ideal mesh. Elements are concentrated at the upper region, and the change from the coarse elements in the lower region to fine elements in the upper region is gradual.

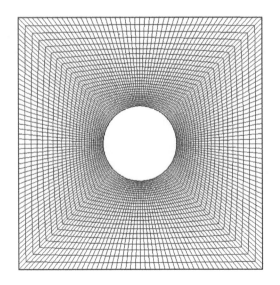

FIGURE 7.9   Mapped mesh of a plate with a hole.

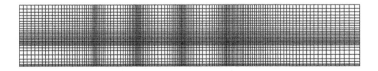

FIGURE 7.10   A mapped mesh of channel containing multiple blocks.

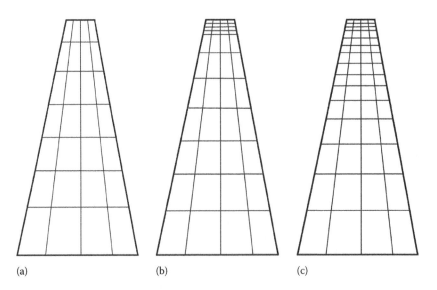

(a)   (b)   (c)

FIGURE 7.11   Different mapped meshes for a fin.

7.4   MAPPED MESH WITH ANSYS

For the geometry shown in Figure 7.12, create a mapped mesh using the ANSYS. The model is intended to be used for a fluid mechanics problem, and therefore, concentrate the mesh at the regions around the step.

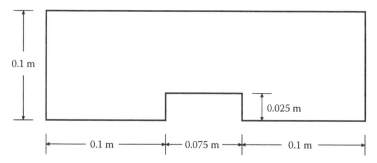

FIGURE 7.12   Geometry for mapped mesh.

Double click on the ANSYS icon

Main Menu > Preferences

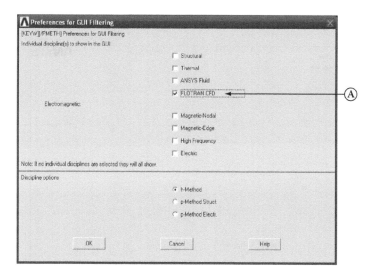

select the Flotran CFD

[OK]

Main Menu > Preprocessor > Element type > Add/Edit/Delete

[Add...]

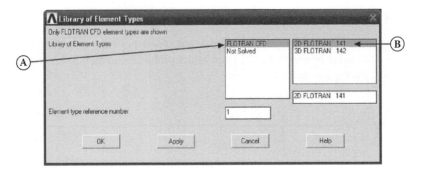

A select FLOTRAN CFD

B 2D FLOTRAN 141

OK

The modeling session is started here. First, 12 key points are created, followed by lines, and finally areas. The *x*- and *y*-coordinates of each key point are identified with its number.

In order to create a mapped mesh with the ANSYS, the computational domain must be divided into areas. Each area must have four sides only. Creating key points that will be connected by line to form four sides area, as follows:

Main Menu > Preprocessor > Modeling > Create > key points > In Active CS

Enter the coordinate of the key points, and the total number of key points is 12, as shown in the figure below.

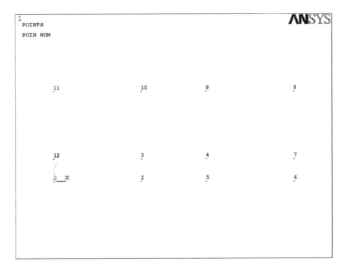

ANSYS graphics shows generated key points

Now, the lines can be created. Using the straight lines, connect two key points to form a line. The connection must be from left to right for all horizontal lines, and for bottom to top for all vertical lines. This technique will make the mesh division easy.

ANSYS Main Menu > Preprocessor > Modeling > Create > Lines > Lines > Straight Line

Click on the Key points to connect two Key points with one straight line, as shown in the figure below.

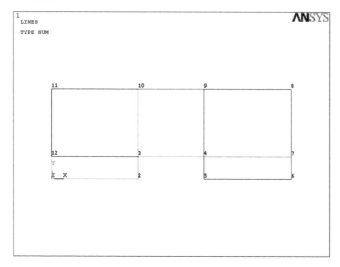

ANSYS graphics shows generated lines

Carefully inspect the above figure; each area is made of four lines. Next, the areas are created.

ANSYS Main Menu > Preprocessor > Modeling > Create > Areas > Arbitrary > By Lines

Click on the lines to create rectangles for the domain as shown in the figure below.

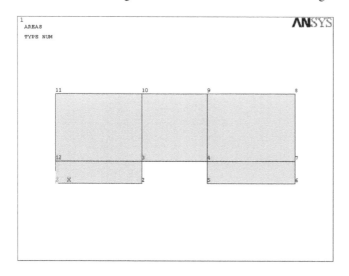

ANSYS graphics shows areas

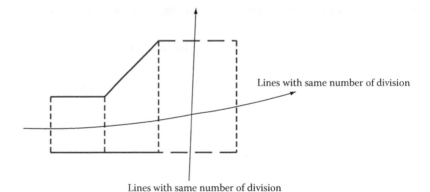

Lines with same number of division

Lines with same number of division

FIGURE 7.13   Mapped mesh.

Now the domain is ready to be meshed. Flow will be separated and reattached at the corners. Elements should be concentrated at the corners since high velocity gradient exists at these locations. To create a mapped mesh, the lines should be divided into segments. The two opposite lines of an area must have same number of divisions. Accordingly, the attached area should have the same number of line divisions. Figure 7.13 shows all opposite lines are having same number of line divisions.

Main Menu > Preprocessor > Meshing > Mesh Tool

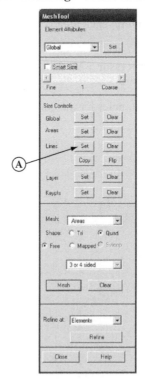

A select Set Line

Select the horizontal lines as shown below.

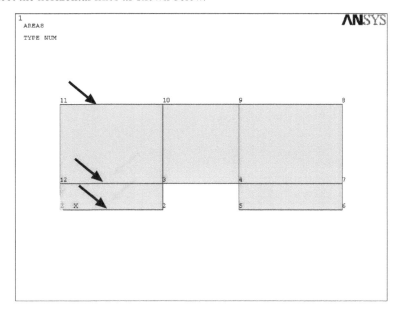

Apply

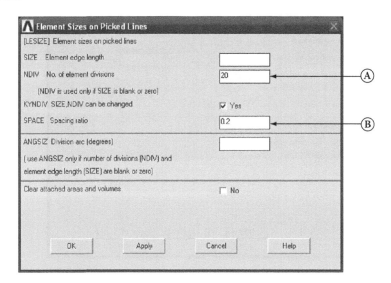

A type 20 in the NDIV

B type 0.2 in the SPACE

The number 20 means that the selected lines will be divided into 20 segments, and the number 0.2 decreases the size of the division by one-fifth.

Apply

Select the horizontal lines as shown below.

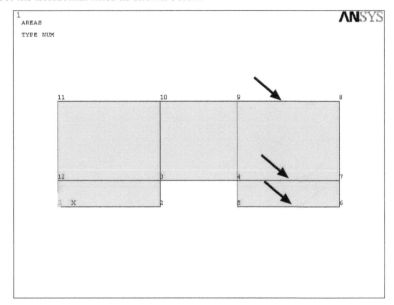

Apply

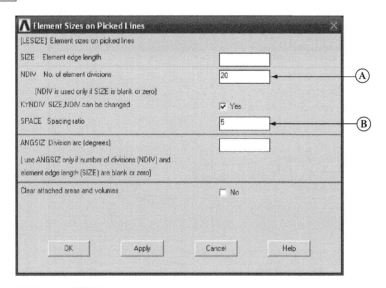

A type 20 in the NDIV

B type 5 in the SPACE

The number 20 means that the selected lines will be divided into 20 segments, and the number 5 increases the size of the division by 5.

Click on the horizontal lines as shown below.

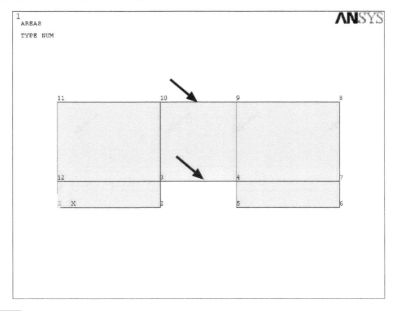

Apply

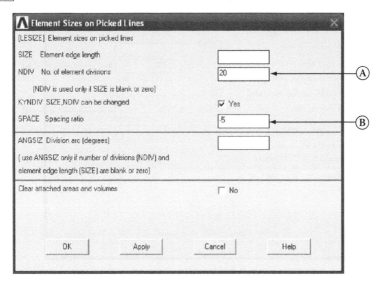

A type 20 in the NDIV

B type −5 in the SPACE

Apply

The number 20 means that the selected lines will be divided into 20 segments, and the number −5 will increase, then decrease the size of the division by 5.

Click on the lines as shown below. These lines will be divided equally.

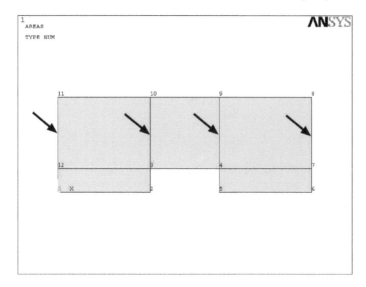

Apply

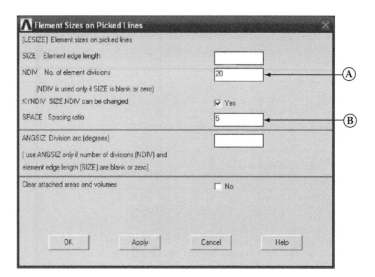

A type 20 in the NDIV

B type 5 in the SPACE

The number 20 means that the selected lines will be divided into 20 segments, and the number 5 increases the size of the division by 5.

Apply

Click on the lines as shown below.

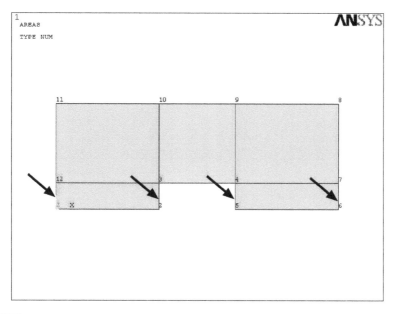

Apply

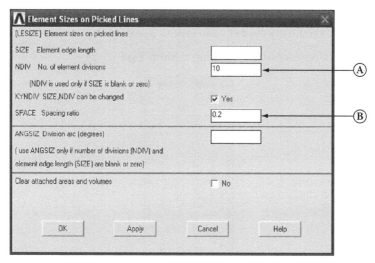

A type 10 in the NDIV

B type 0.2 in the SPACE

The number 10 means that the selected lines will be divided into 20 segments, and the number 0.2 decreases the size of the division by one-fifth. ANSYS graphics shows the lines have been meshed according to the specified mesh divisions.

OK

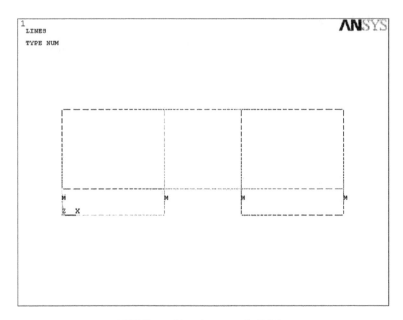

ANSYS graphics shows mesh division

Inspect the mesh carefully; all opposite sides have the same number of divisions; otherwise an error message will appear or the mesh will not be mapped. Now the mesh is ready to be meshed. The Smart Size selection must be off, keep the element shape as Quad, and finally change mesh type to Mapped.

Main Menu > Preprocessor > Meshing > Mesh Tool

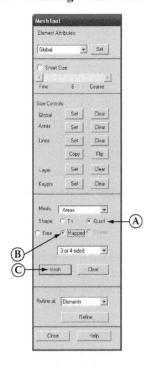

A select Quad

B select Mapped

C click on Mesh

| Pick All |

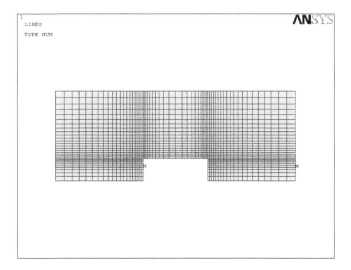

ANSYS graphics shows a mapped mesh

PROBLEMS

7.1 Create a mapped mesh for the geometries shown in Figure 7.14, and the geometries are used for solid mechanics.

7.2 Create a mapped mesh for the geometries shown in Figure 7.15, and the geometries are used for heat transfer.

7.3 Create a mapped mesh for the geometries shown in Figure 7.16, and the geometries are used for fluid mechanics.

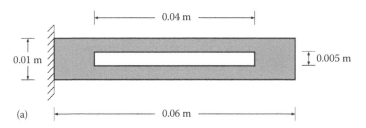

FIGURE 7.14   Geometries for solid mechanics.

(*continued*)

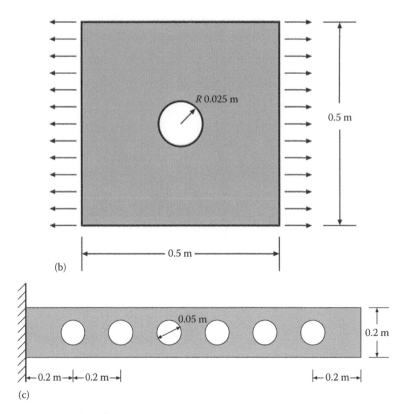

(b)

(c)

FIGURE 7.14 (continued)

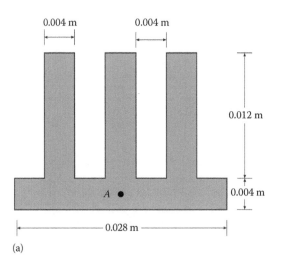

(a)

FIGURE 7.15   Geometries for heat transfer.

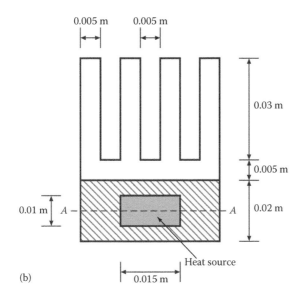

(b)

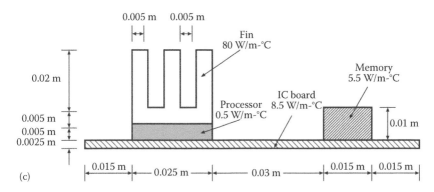

(c)

FIGURE 7.15 (continued)

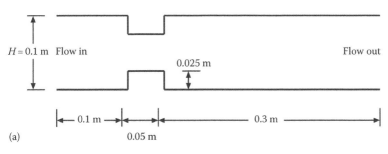

(a)

FIGURE 7.16 Geometries for fluid mechanics.

(*continued*)

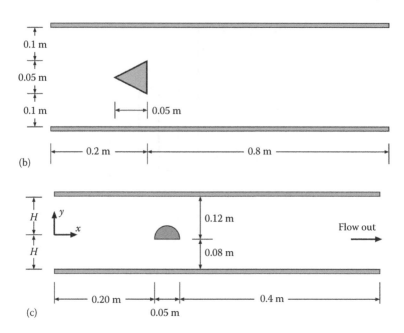

FIGURE 7.16 (continued)

Bibliography

V. Adams and A. Askenazi. *Building Better Product with Finite Element Analysis*. ONWORD Press, Santa Fe, NM, 1999.

M. Fagan. *Finite Element Analysis: Theory and Practice*. Pearson Education Limited, Harlow, Essex, England, 1992.

S. Moaveni. *Finite Element Analysis: Theory and Application with ANSYS*. Prentice Hall, Upper Saddle River, NJ, 2007.

T. Chandrupatla and A. Belegundu. *Introduction to Finite Elements in Engineering*, 3rd edn. Prentice Hall, Upper Saddle River, NJ, 2002.

J. Reddy. *An Introduction to the Finite Element Method*, 2nd edn. McGraw Hill, New York, 1993.

E. Thompson. *Introduction to the Finite Element Method: Theory, Programming, and Applications*. John Wiley & Sons, Hoboken, NJ, 2005.

ANSYS Theory, ANSYS online manual.

Index